W0036576

Synthesis Lectures on Computer Vision

Series Editors

Gerard Medioni, University of Southern California, Los Angeles, CA, USA

Sven Dickinson, Department of Computer Science, University of Toronto, Toronto, ON, Canada

This series publishes on topics pertaining to computer vision and pattern recognition. The scope follows the purview of premier computer science conferences, and includes the science of scene reconstruction, event detection, video tracking, object recognition, 3D pose estimation, learning, indexing, motion estimation, and image restoration. As a scientific discipline, computer vision is concerned with the theory behind artificial systems that extract information from images. The image data can take many forms, such as video sequences, views from multiple cameras, or multi-dimensional data from a medical scanner. As a technological discipline, computer vision seeks to apply its theories and models for the construction of computer vision systems, such as those in self-driving cars/navigation systems, medical image analysis, and industrial robots.

Terrance Boult · Walter Scheirer
Editors

A Unifying Framework for Formal Theories of Novelty

Discussions, Guidelines, and Examples for Artificial Intelligence

 Springer

Editors
Terrance Boult
Department of Computer Science
University of Colorado at Colorado Springs
Colorado Springs, CO, USA

Walter Scheirer
Department of Computer Science
and Engineering
University of Notre Dame
Notre Dame, IN, USA

ISSN 2153-1056   ISSN 2153-1064   (electronic)
Synthesis Lectures on Computer Vision
ISBN 978-3-031-33053-7   ISBN 978-3-031-33054-4   (eBook)
https://doi.org/10.1007/978-3-031-33054-4

© The Editor(s) (if applicable) and The Author(s), under exclusive license to Springer Nature
Switzerland AG 2024

This work is subject to copyright. All rights are solely and exclusively licensed by the Publisher, whether the whole
or part of the material is concerned, specifically the rights of translation, reprinting, reuse of illustrations, recitation,
broadcasting, reproduction on microfilms or in any other physical way, and transmission or information storage
and retrieval, electronic adaptation, computer software, or by similar or dissimilar methodology now known or
hereafter developed.
The use of general descriptive names, registered names, trademarks, service marks, etc. in this publication does
not imply, even in the absence of a specific statement, that such names are exempt from the relevant protective
laws and regulations and therefore free for general use.
The publisher, the authors, and the editors are safe to assume that the advice and information in this book are
believed to be true and accurate at the date of publication. Neither the publisher nor the authors or the editors give
a warranty, expressed or implied, with respect to the material contained herein or for any errors or omissions that
may have been made. The publisher remains neutral with regard to jurisdictional claims in published maps and
institutional affiliations.

This Springer imprint is published by the registered company Springer Nature Switzerland AG
The registered company address is: Gewerbestrasse 11, 6330 Cham, Switzerland

Preface: The Novelty Problem in AI

AI researchers these days are quick to tout the progress that has been made in the field over the past decade. From game playing to visual recognition, new capabilities are appearing all of the time for many different applications. And indeed, such achievements should be celebrated. However, some very useful AI capabilities remain out of reach. For instance, why aren't safe self-driving cars available in the market in 2023? A major limitation of today's AI systems has become apparent in the quest for autonomous systems that must operate in real environments: they cannot manage novelty in the environment they were designed for. That is to say, if something new appears, there is no capacity for an agent to detect, characterize, and learn how to handle it. Given the practically infinite number of ways an environment can configure itself, coupled with the routine appearance of new things within an environment, novelty can be a significant confound. This book is the first attempt to study novelty problems in a rigorous fashion through the use of a unifying framework for formal theories of novelty.

Ad hoc ways of addressing the novelty problem have proven to be insufficient. There exists a persistent belief that reinforcement learning is all that is needed for agents to manage novelty because any novelty can be learned over time. Similarly, there exists blind faith in the generalization properties of deep learning through invariant representation alone. Given the results found in this book for the simplest of AI domains, we can safely say that a more principled approach to novelty management is needed. Neither approach addresses the core detection problem at the classifier level, nor is there any capacity to characterize novelty, which can take on many different forms. On that latter point, what exactly does it mean for something to be novel? That isn't a question that can be answered using an off-the-shelf AI algorithm. The need for a theory matched to a specific domain can provide a better starting point for agent design.

A recent effort to address the novelty problem in AI has been the DARPA Science of Artificial Intelligence and Learning for Open-world Novelty (SAIL-ON) program. It established a research program to develop a set of engineering design principles for open-world learning in 2019 [1]. Throughout the four years of the program, a large consortium of academic, industry, and government researchers has collaborated on fundamental work looking into innovative strategies for effective open-world learning in both activity domains, e.g., interactive video games, and perceptual domains, e.g., datasets of images and videos. This book is the output of the program's Novelty Working Group, which was charged with developing viable theories for the study of novelty in AI. Each chapter was contributed by different participants in that working group.

This book is organized in the following manner. Chapter 1 is the focal point of the book. It introduces a unifying framework for creating theories of novelty that are matched to specific domains. This includes definitions on different types of novelty, as well as constructs for building agents that can detect, characterize, and manage novelty. This framework is general and can apply to *any* domain in which novelty appears. To justify this claim, each subsequent chapter provides an example domain for which a theory is developed and evaluated. These chapters include: (1) a task overview, (2) definitions of dissimilarity and regret operators, (3) definitions of measurements and observations, (4) a description of novelty types and examples, (5) a set of experiments validating predictions made by the developed theory, and (6) concluding remarks.

The domain-specific chapters cover a broad range of activity and perceptual domains. Chapter 2 starts things off as simple as possible with a study of novelty in the 2D Cart-Pole activity domain. Chapter 3 extends the study of CartPole by examining a 3D version of the environment. Chapter 4 turns to the perceptual domain of image classification in computer vision. Chapter 5 discusses a related computer vision domain, handwriting recognition, which also contains elements of natural language processing. Chapter 6 pushes farther into the realm of natural language processing by studying contextual and semantic novelty in text. Chapter 7 comes back to activity domains with an examination of the game Monopoly. The book concludes in Chap. 8 by recapping what we have learned and suggesting the development of new theoretical directions that are interdisciplinary in nature.

Colorado Springs, USA   Terrance Boult
Notre Dame, USA   Walter Scheirer

Acknowledgments This research was sponsored by the Defense Advanced Research Projects Agency (DARPA) and the Army Research Office (ARO) under multiple contracts/agreements including HR001120C0055, W911NF-20-2-0005, W911NF-20-2-0004, HQ0034-19-D-0001, and

W911NF2020009. The views contained in this document are those of the authors and should not be interpreted as representing the official policies, either expressed or implied, of DARPA or ARO, or the US Government.

Reference

1. DARPA. Teaching AI systems to adapt to dynamic environments, (2019)

Contents

List of Figures

A Unifying Framework for Novelty

1

T. Boult, D. S. Prijatelj and W. Scheirer

1.1   Introduction

"What is novel?" is an important AI research question that informs the design of agents tolerant to novel inputs. Is a noticeable change in the world that does not impact an agent's task performance a novelty? How about a change that impacts performance but is not directly perceptible? If the world has not changed, but the agent senses a random error that produces an input that leads to an unexpected state, is that novel?

With decades of work and thousands of papers covering novelty detection and related research in anomaly detection [1], out-of-distribution detection, open-set recognition, and open-world recognition, one would think that a consistent unified definition of novelty would have been developed. Unfortunately, that is not the case. Instead, we find a plethora of variations on this theme, as well as ad hoc use and inconsistent reuse of terminology, all of which injects confusion as researchers discuss these topics.

This chapter introduces a unifying formal framework of novelty. The framework seeks to formalize what it means for an input to be a novelty in the context of agents in artificial intelligence or in other learning-based systems. Using the proposed framework, we formally define multiple types of novelty an agent can encounter. The goal of these definitions is to be broad enough to encompass and unify the full range of novelty models that have been proposed in the literature [2–7]. An important generalization beyond prior work is that we consider novelty in the world, observed space, and agent space (see Fig. 1.1), with

T. Boult (✉)
University of Colorado Colorado Springs, Colorado Springs, CO, USA
e-mail: tboult@vast.uccs.edu

D. S. Prijatelj · W. Scheirer
University of Notre Dame, Notre Dame, IN, USA

© The Author(s), under exclusive license to Springer Nature Switzerland AG 2024
T. Boult and W. Scheirer (eds.), *A Unifying Framework for Formal Theories of Novelty*, Synthesis Lectures on Computer Vision,
https://doi.org/10.1007/978-3-031-33054-4_1

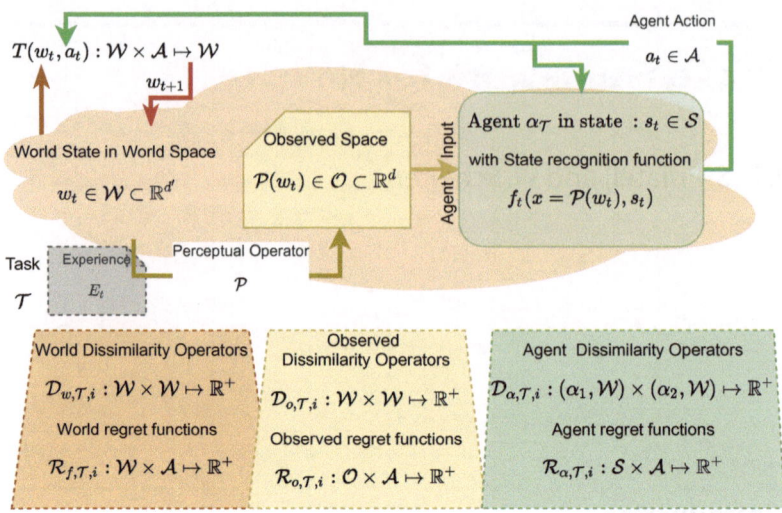

Fig. 1.1 Main elements of the implicit theories of novelty; items with dashed outlines are outside of the task or agent but are critical to defining novelty. In the framework, a theory of novelty is obtained by specifying: world $\mathcal{W}$ and the world dimensionality d', observation space $\mathcal{O}$ accessible to the agent, and agent space $\mathcal{S}$. The agent can only access world information indirectly through a perceptual operator $\mathcal{P}$ with associated $\mathcal{W}_t$ processed world regions. The agent α in state $s_t \in \mathcal{S}$ at time t, using state recognition function $f_t(x, s)$ to determine the action $a_t \in \mathcal{A}$ to be taken, which is used by an oracle's world state transition function T. Critical to defining novelty are task-dependent world dissimilarity functions $\mathcal{D}_{w,\mathcal{T};E_t}$ with associated threshold δ_w, task-dependent observation space dissimilarity functions $\mathcal{D}_{o,\mathcal{T};E_t}$ with threshold δ_o, world regret function $\mathcal{R}_{w,\mathcal{T}}$, observation space regret function $\mathcal{R}_{o,\mathcal{T}}$, and agent space regret function $\mathcal{R}_{a,\mathcal{T}}$. The framework also defines a task-dependent agent space dissimilarity function $\mathcal{D}_{\alpha,\mathcal{T};E_t}$ which allows one to consider the difference between agent models for different worlds, *e.g.*, to measure learning. While not explicit in the figure, the world state and observed states may collapse, if perceptual operator is the identity. In addition, the world and observed spaces may include full or partial copies of the agent's memory/state/state recognition function. Every set of these operators/functions/values defines a different theory of novelty for its associated task

dissimilarity and regret operators critical to our definitions. The overarching goal is a framework such that researchers have clear definitions for the development of agents that must handle novelty, including support for agents/algorithms that incrementally learn from novel inputs.

Our framework supports *implicit theories of novelty*, meaning the definitions use functions to implicitly specify if something is novel. The framework does not require a way to generate novelties, but rather it provides functions that can be used to evaluate if a given input is novel. This is similar to how any 2D shape can be implicitly defined by a function $f(x, y) = 0$, whether or not there is a procedure for generating the shape. We contend any constructive or generative theory of novelty [7] must be incomplete because the construction or generation of defined worlds, states, and any enumerable sets of transformation between them form, by definition, a closed-world. We note, however, that a constructive model can be consistent with our definition, but we do not require a constructive model.

References

1. Gruhl C, Sick B, Tomforde S (2021) Novelty detection in continuously changing environments. Future Gener Comput Syst 114:138–154
2. Pimentel MAF, Clifton DA, Clifton L, Tarassenko L (2014) A review of novelty detection. Signal Process 99:215–249
3. Markou M, Singh S (2003) Novelty detection: a review-part 1: statistical approaches. Signal Process 83(12):2481–2497
4. Markou M, Singh S (2003) Novelty detection: a review-part 2: neural network based approaches. Signal Process 83(12):2499–2521
5. Scheirer WJ, de Rezende Rocha A, Sapkota A, Boult TE (2013) Toward open set recognition. IEEE TPAMI 35(7):1757–1772
6. Bendale A, Boult TE (2015) Towards open world recognition. In: The IEEE conference on computer vision and pattern recognition (CVPR), pp 1893–1902
7. Langley P (2020) Open-world learning for radically autonomous agents. In: AAAI. JSTOR, pp 13539–13543

Novelty in 2D CartPole Domain

P. A. Grabowicz, C. Pereyda, K. Clary, R. Stern, T. Boult, D. Jensen
and L. B. Holder

In this chapter, we apply the constructs defined in Chap. 1 to the relatively simple domain of 2D CartPole. We consider different versions of CartPole agents to highlight the different effects of novelties on an agent. The results generally show that the novelty theoretic framework allows the estimation of different novelties' impact on the performance of agents, which can be used for testing open-world learning hypotheses. This chapter presents a few surprising results that help further inform the framework.

2.1 Task Overview

As an example of using the constructs defined above, consider the CartPole game in the OpenAI Gym.[1] In this domain, a cart has a pole connected to it, and the task $\mathcal{T}$ is to push the cart left or right so as to prevent the attached pole from falling. The world state w in the CartPole domain comprises the following real values (with default/initial values in paren-

---

[1] https://github.com/openai/gym/blob/master/gym/envs/classic_control/cartpole.py.

P. A. Grabowicz (✉) · K. Clary · D. Jensen
University of Massachusetts Amherst, Amherst, MA, USA
e-mail: grabowicz@cs.umass.edu

C. Pereyda · L. B. Holder
Washington State University, Pullman, WA, USA

R. Stern
PARC, Palo Alto, CA, USA

T. Boult
University of Colorado Colorado Springs, Colorado Springs, CO, USA

© The Author(s), under exclusive license to Springer Nature Switzerland AG 2024
T. Boult and W. Scheirer (eds.), *A Unifying Framework for Formal Theories
of Novelty*, Synthesis Lectures on Computer Vision,
https://doi.org/10.1007/978-3-031-33054-4_2

theses): gravity G (9.8), mass of cart M_c (1.0), mass of pole per unit length M_p (0.1), length of pole L (1.0), force of push F_p (10.0), horizontal force acting on the cart F_h (0), min/max cart position z^{min} (-2.4), z^{max} ($+2.4$), min/max pole angle ϕ^{min} ($-12°$), ϕ^{max} ($+12°$), time between state updates τ (0.02 seconds), and start time t (0). The initial cart position z_0, cart velocity $\dot{x}_0$, pole angle ϕ_0, and pole angular velocity $\dot{\phi}_0$ are all i.i.d. random samples from $[-0.05 \ldots 0.05]$. The perceptual operator $\mathcal{P}$ in this domain is a projection of the world state that returns only the cart position z, cart velocity $\dot{z}$, pole angle ϕ, and pole angular velocity $\dot{\phi}$ as the 4D observed state vector $x = (z, \dot{z}, \phi, \dot{\phi})$. Overall, each world state contains the parameters, observations x, and state transition and perceptual operators, that is $w = (G, M_c, M_p, L, F_p, F_h, z^{min}, z^{max}, \phi^{min}, \phi^{max}, \tau, x, \mathcal{P}, T)$.

The task is more precisely defined as: given x, select an action from the space of $\mathcal{A} = \{$Left, Right$\}$ to maintain the cart position within the min/max cart position and maintain the pole angle within the min/max pole angle as long as possible. State transitions in the CartPole domain are determined by the equations of motion and the action chosen by the agent, and can be simulated in discrete time with numerical integration (see Florian [1] for details). The agent's task is to keep the pole upright as long as possible. Each CartPole episode lasts up to $t_{max} = 200$. Each episode ends when the agent fails to keep the pole upright, as defined above, or when $t = t_{max}$, whichever comes first. In this example, we assume that transitions between observed states are Markovian, which simplifies presentation; other theories for this domain could consider dissimilarity and regret in more general settings.

2.2   Dissimilarity and Regret

2.2.1   Dissimilarity Measures

One approach to introducing a dissimilarity measure is based on the normalized difference between world states w_1 and w_2, with the assumption that an agent is trained in world w_1 but then evaluated in world w_2. For example, world w_1 may use a pole length of one meter, and world w_2 uses a pole length of two meters. The world difference-based dissimilarity $\tilde{\mathcal{D}}^d_{w,T}$ for task T is defined as,

$$\tilde{\mathcal{D}}^d_{w,T}(w_1, w_2) = \sum_{i=1}^{10} \frac{w_{2,i} - w_{1,i}}{\max(|w_{2,i} - w_{1,i}|)}. \tag{2.1}$$

The normalization $\tilde{\mathcal{D}}$ occurs separately for the i'th dimension of the world state. The expression sums the individual $(w_{2,i} - w_{1,i})$ divided by the maximum possible range of changes to the i'th dimension between the two worlds. In this chapter, we only apply changes to one parameter at a time, so both the enumerator and denominator are zeroes for every dimension except one, and we replace these singularities with zeroes. Note that the difference in the numerator is not squared or an absolute value. This differs from the novelty framework in which dissimilarity is assumed non-negative. However, in the case of CartPole and likely other domains, the direction of change in the world state impacts the performance of an agent in the novel world. For example, an agent trained on a higher gravity than in the novel world may perform better than an agent trained on a lower gravity.

The non-negative dissimilarity measure for CartPole might take a simple form, e.g., the Euclidean distance in the world or observed space. However, Euclidean distance between world states is affected by factors other than novelties, including the choice of units. It is also insensitive to the variation in the impact of different variables on the state evolution or task outcome. Proper operationalization would reduce dependency on units and account for states that correspond to different samples from the same world, e.g., the same CartPole world with a different initial position of the pole is not considered novel.

To avoid these issues, we compare the two world states, w and $\breve{w}$, by comparing the states proceeding from a common observed state and action in each of the two worlds. We consider an optimal agent's action, a^*, that is optimal in the first world w, and choose as the common observed states the states that the agent encounters, $\check{x}$, in the second world, $\breve{w}$,

$$\mathcal{D}_{o,\mathcal{T}}(w, \breve{w}) = (\mathcal{P}(T(M(w, \check{x}), a^*)) - \breve{\mathcal{P}}(\breve{T}(M(\breve{w}, \check{x}), a^*)))^2,$$

where $M(w, x) : \mathcal{W} \mapsto \mathcal{W}$ is a function that returns a modified w whose observed components are replaced with the values from x, such that $\mathcal{P}(M(w, x)) = x$, while all other components remain unchanged, i.e., the function intervenes on the world state. Note that the operators T and $\mathcal{P}$ could be affected by novelty and, hence, they could differ between w and $\breve{w}$. The proposed measure captures any meaningful novelties in the operators by using the operators that correspond to their world, e.g., the operators can be represented as programs whose code is contained in the respective world states.

Overall, the dissimilarity measures the distance between observed states in two different worlds that proceed from a common observed state and action, $\check{x}$ and a^*. The agent is optimal in the first world, while the trajectory is from the second world, so this dissimilarity measure can be seen as a state prediction error of the optimal agent trained in the first world and tested in the second world. The world dissimilarity is defined analogously, except it does not use the perceptual operator,

$$\mathcal{D}_{w,\mathcal{T}}(w, \breve{w}) = (T(M(w, \check{x}), a^*) - \breve{T}(M(\breve{w}, \check{x}), a^*))^2,$$

so it captures the differences between unobserved parts of the world states.

Note, this is an asymmetric dissimilarity measure as the selected agent is optimal for the first world and need not be optimal for the second. Due to the conditioning, any pair of states from the same world will have zero dissimilarity. Furthermore, since these depend on the choice of "optimal action" a^* from the first world, it implicitly normalizes for how different dimensions (variables) impact the evolution of the world state. One can consider actions of an optimal reference agent under given conditions, e.g., in certain conditions a simulation-based one-step lookahead agent will perform nearly optimally in a non-novel

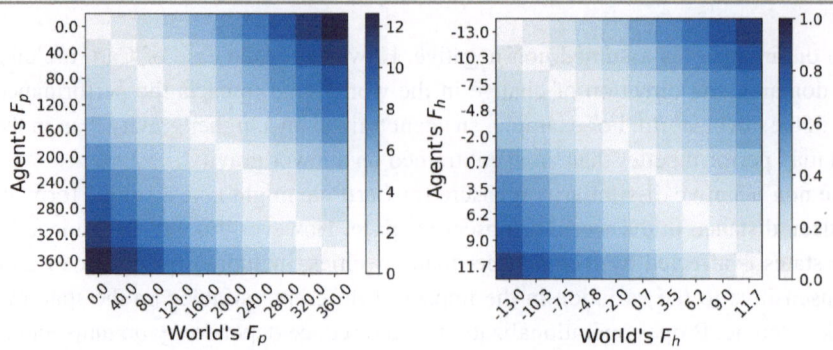

Fig. 2.1 Average dissimilarity, $\mathbb{E}_{\breve{w}} \mathcal{D}_{o,\mathcal{T}}(w_t, \breve{w}_t)$, between the future states expected and observed by agents that were tuned to the world w with incorrect value of the magnitude of pushing force, F_p (left panel), or a horizontal force acting on the cart, F_h (right panel), while tested in the world $\breve{w}$. The expectation is computed over 20 samples of the initial world state w_0

CartPole environment, whereas in other conditions a Deep Q-learning Network (DQN) agent will perform nearly optimally. If it is infeasible to obtain an optimal agent in practice, then one may use an arbitrary reference agent and its expectation, with the caveat the dissimilarity measure will depend on the oracle's reference agent.

The dissimilarity measure quantifies the difference between world states at a given time step. We measure the average dissimilarity over agent's trajectory in the test world, $\breve{w} = \{\breve{x}_t\}$, i.e., over time and initial conditions. As expected, the average dissimilarities in observed state prediction captures the novelties in the magnitude of pushing force and in a horizontal force acting on the cart (Fig. 2.1). The larger the distance between the value of a given parameter in the environment and its value assumed by the agent, the larger the dissimilarity in state prediction. Hence, the average observed dissimilarity measure can be used by a novelty-aware agent to detect novelty.

2.2.2 Regret

Regret for the agent's action at state x_t w.r.t. $\mathcal{T}$ is,

$$\mathcal{R}_{w,\mathcal{T}}(w_t, a_t) = \ell_{\mathcal{T}}(w_t, a_t) - \ell_{\mathcal{T}}(w_t, a_t^{*,w_t}),$$

where $\ell_{\mathcal{T}}(w_t, a_t)$ is the loss incurred in the world w_t by a given agent that observes state x_t and performs action a_t, while $\ell_{\mathcal{T}}(w_t, a_t^{*,w_t})$ is the loss of an agent that performs the optimal action a_t^{*,w_t}, given the knowledge of world state w_t, including its unobserved part. In CartPole, the loss at the given time step is 1 if the pole angle or cart's position in the next time steps are beyond the threshold, $|\phi_{t+1}| > \phi_{t+1}^{max}$ or $|z_{t+1}| > z_{t+1}^{max}$, otherwise the loss is 0. If at time t the loss is 1, then the episode ends, the agent stops taking actions, and the loss $\ell_{\mathcal{T}}(w_{t'}, \cdot) = 1$ for all consecutive time steps $t' \in \{t + 1, \ldots, t_{max}\}$.

The observation regret is the same as the world regret,

$$\mathcal{R}_{o,\mathcal{T}}(w_t, a_t) = \ell_{\mathcal{T}}(w_t, a_t) - \ell_{\mathcal{T}}(w_t, a_t^*),$$

except that the optimal agent that takes the action a_t^* observes only x_t and does not have any knowledge of the unobserved part of w_t. The two regret measures computed for the same world state w_t may differ, because the optimal action may depend on the agent's knowledge of the hidden part of w_t. For instance, if there are some hidden dynamic elements interacting with pole or cart, such as an invisible pendulum that hangs above the pole and sometimes hits it, knowledge of the state of that hidden element may affect the optimal decision. There are no such hidden elements in the default CartPole environment, but they could be introduced as a novelty. Despite this, the observation regret depends on the world state and not the observed state, because the state transition function acts on the world state to compute the next state and to determine the loss associated to the agent's action.

2.3   Measurements and Observations

The performance-based measurements in the CartPole domain have already been introduced, and are based on the number of time steps, $t_\alpha \in \{0, \ldots, t_{max}\}$ that the agent can keep the pole balanced. To normalize this measurement, we use the maximum time t_{max} for each episode of the CartPole game, so that performance is t_α/t_{max}. Based upon this performance measure, we define the expected loss as,

$$\mathcal{L}(\alpha) = \mathbb{E}_{w,a}\left[\ell(w_t, a_t)\right], \tag{2.2}$$

which averages over initial and proceeding states, $w = \{w_t\}$, as well as agent's actions, $a = \{a_t\}$. The expected loss stands for the reduction in performance as a result of the novelty with respect to the maximal performance; however, under certain conditions the maximal performance may be unachievable by any agent due to the physical constraints of the system. If agent α's actions for all world states are deterministic, then we can write $\mathbb{E}_{w,a}\left[\ell(w_t, a_t)\right] = \mathbb{E}_w\left[\sum_{t=0}^{t_{max}} \ell(w_t, a_t)/t_{max}\right]$ and the expectation is only over the random initial state w_0 and possibly stochastic state transitions.

In addition, we measure agent's regret $R(\alpha, \alpha^*)$, which is the difference in expected loss between the target agent and an optimal agent evaluated in the same world,

$$R(\alpha, \alpha^*) = \mathcal{L}(\alpha) - \mathcal{L}(\alpha^*). \tag{2.3}$$

Here, for simplicity, we take as the optimal agent the same agent as the target agent, the difference being that the optimal agent, α^*, is tuned or trained in the test environment, $\breve{w}$, hence we refer to it as $\breve{\alpha}$, whereas the target agent, α, is tuned or trained in a potentially different training environment, w.

In general, measurements can be defined based on the differences between the worlds in which the agent is trained, w, and tested, $\check{w}$. In the following experiments, we define the CartPole world by a set of real-value parameters, e.g., gravity, pole length, and the measurement of the normalized difference-based dissimilarity in these parameter values, $\tilde{\mathcal{D}}^d_{w,\mathcal{T}}(w, \check{w})$.

Another type of measurement that we do not consider in this chapter is the measurements that are suitable for novelty detection. One simple approach to developing such measures is by computing the expected error between the agent's prediction of the next observable state and the actual value of the state, $\mathcal{D}_{o,\mathcal{T}}(w, \check{w})$. A significant increase in prediction error may signal the presence of novelty. In CartPole, the observable sensors are a small subset of the parameters that describe the world. In fact, multiple different novelties can result in similar deviations to observable sensors, e.g., decrease in push force versus increase in cart mass. The measurement of novelty detection is considered in more detail in the next chapter on a 3D version of the CartPole domain.

2.4   Novelty Types and Examples

The introduced notions of dissimilarity and regret can be used to categorize novelty types in the CartPole environment. The novelties in the pushing force and a horizontal force acting on the cart, introduced in Fig. 2.1, are both world and observation novelties, since both world and perceptual dissimilarities are larger than zero, $\mathcal{D}_{o,\mathcal{T}}(\check{w}_t, w_t) > 0$ and $\mathcal{D}_{w,\mathcal{T}}(\check{w}_t, w_t) > 0$.

If the CartPole environment had an additional unobserved cart that does not influence the main cart and pole, then any novelty in that unobserved cart would be an imperceptible world novelty since it would not influence transitions between the observed states, $\mathcal{D}_{o,\mathcal{T}}(\check{w}_t, w_t) = 0$ and $\mathcal{D}_{w,\mathcal{T}}(\check{w}_t, w_t) > 0$ and would be categorized as an imperceptible nuisance novelty.

2.5   Experiments

To evaluate the application of the novelty theory framework in the 2D CartPole domain, we consider the performance of two different agents across four different observation novelties: changes in push force F_p, horizontal force applied to the cart F_h, gravity G, and pole length L. The first agent is a simple simulation-based lookahead agent whose parameters are tuned to the non-novel environment. The second agent is a DQN agent that is trained on the non-novel environment but then is held fixed during testing.

2.5.1   Simulation-Based Agent

A particularly simple version of CartPole is the one where the agent's action space is binary, i.e., the agent can choose to push the cart left or right, and its reward is the time that the pole is up. In such a CartPole environment, an agent can often find nearly-optimal actions by performing what-if simulations of the world and searching for actions that result in the best performance, i.e., the agent can simulate what would happen if it pushed the cart left or right and then chose the next action that results in better performance. In this paper, for simplicity, we present the results for a simulation-based single lookahead agent. The action is chosen based on the distance, $||\beta^{\mathsf{T}}(x_t - x^s)||$, between the expected state resulting from the action, x_t, and the desired state, $x^s = (0, 0, 0, 0)$, where the weight vector $\beta = (0, 0, 1, 0.005)$ weighs discrepancy in ϕ the most and ignores the discrepancies in z and $\dot{z}$. The state of this agent is described by the parameters used to simulate system dynamics and the weight vector, $s = (G, M_c, M_p, L, F_p, F_h, \tau, \beta)$. All the parameters, with the exception of the action-selection parameter β that does not have a correspondence in the environment, are tuned manually to the values that either correspond to the environment (no novelty) or do not (novelty). For simplicity, we set the parameters manually to the desired values, but they could be learned, e.g., via Monte Carlo sampling. If the learning was performed in the test environment, then we would refer to this agent as adaptive.

2.5.1.1 Results

Next we vary the world parameters of pole length L, gravity G, push force F_p, and horizontal force F_h in both w and $\breve{w}$ to observe the extent of the impact on performance as a function of the difference-based dissimilarity between w and $\breve{w}$, $\tilde{\mathcal{D}}^d_{w,\mathcal{T}}(w, \breve{w})$. The results are presented in three forms in Fig. 2.2. On the left are heatmaps showing the normalized performance difference for the different pairs of parameters, e.g., $(L, \breve{L})$. The middle and right plots show the normalized expected loss $\mathcal{L}(\alpha)$ and regret $\mathcal{R}(\alpha, \breve{\alpha})$ of a simulation-based agent tuned in the non-novel world and tested in the novel world, α, compared with the same agent tuned and tested in the novel world, $\breve{\alpha}$. These two measures are plotted against the difference-based normalized dissimilarity $\tilde{\mathcal{D}}^d_{w,\mathcal{T}}$, which for our single-parameter-varying experiments is just the normalized difference between respective parameter's value in the training world w and test world $\breve{w}$. Note that the expected regret is the average over the loss values at a certain distance from the heatmap's diagonal minus the loss value on the diagonal; hence the regret is zero for $\tilde{\mathcal{D}}^d_{w,\mathcal{T}} = 0$.

To distinguish between novelties that affect or do not affect the task performance, we can use the expected regret, $\mathcal{R}(\alpha, \breve{\alpha})$. Novelties with $\mathcal{R}(\alpha, \breve{\alpha}) = 0$ do not affect task performance and can be ignored. By contrast, $\mathcal{R}(\alpha, \breve{\alpha}) > 0$ tells us that the novelty impacts agent's performance and that the agent can update its state to improve its performance, i.e., learn the novelty. Naturally, one can develop an adaptive version of this agent by learning an estimate of world state parameters from observations and using them to perform more accurate simulations and taking better actions.

Surprisingly, in CartPole, reasonable changes to some of the aforementioned latent parameters do not impact a simulation-based non-adaptive agent's performance (Fig. 2.2). For instance, any changes to the magnitude of pushing force do not affect that agent's perfor-

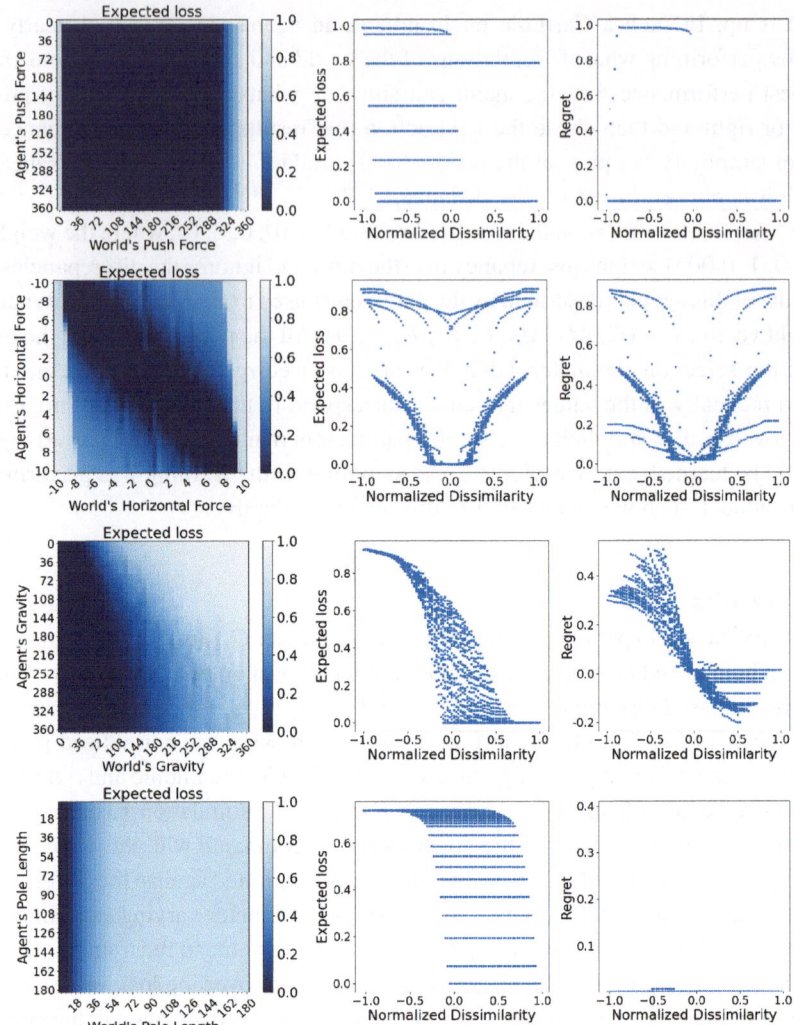

Fig. 2.2 Expected loss and regret of simulation-based agent when varying: **a** *push force* from 0 to 360 N, **b** a constant *horizontal force* from −10 to +10 N, **c** *gravity* from 0 to 360 m/s², **d** *pole length* from near 0 to 180 m. The heatmap (left) shows the difference in performance between an agent trained in agent's world, w, and evaluated on a different test world, $\breve{w}$. The scatter plots show the expected loss (middle) and regret (right) in the novel environment as a function of difference-based dissimilarity, $\breve{\mathcal{D}}_{w,\mathcal{T}}^{d}(w, \breve{w})$

mance (row 1 of Fig. 2.2), even if the agent's simulation uses a wrong value of that parameter. The sudden performance drops on the verges of the respective heatmap correspond to the achievable limits in performance, i.e., the left verge corresponds to such a small force magnitude that the push is insufficient to counter the gravity, whereas the right verge corresponds to such a large force magnitude that the push instantly rotates the pole beyond the allowed region (a theoretical analysis estimating these limits is provided in Sect. 2.5.3). We conclude that the simulation-based agent performs just as well in the novel world, despite making simulations that assume incorrect values of pushing force, i.e., the values from the non-novel world. Note that this agent is non-adaptive, so it does not change its internal model to adapt to the new environment, i.e., it has exactly the same internal model as in the non-novel environment.

Similarly, small changes to the horizontal force (changes up to 3 N, which correspond to $|\tilde{\mathcal{D}}_{w,\mathcal{T}}^d(w, \check{w})| < 0.3$) do not affect that agent's performance (row 2 of Fig. 2.2). Again, the performance drops at the verge of the heatmap mark inherent performance limits, i.e., if the horizontal force is larger than the magnitude of the pushing force, $F_p = 10$ N, then the pushes are insufficient to counteract the horizontal force and the pole is destined to fall, despite taking optimal actions.

Certain latent parameters impact the agent's performance in an intuitive way, i.e., the more their value differs from agent's value, the larger the expected loss of the agent in the test world, e.g., changes to the aforementioned horizontal force result in this behavior (row 2 in Fig. 2.2). The effect is symmetric with respect to the diagonal of the heatmap. The symmetry is also visible in the expected loss and regret, since the dissimilarity corresponds to the distance to the diagonal.

More surprisingly, other novelties impact performance in less intuitive ways, e.g., changes to the gravity have barely any effect if the gravity in the test world is kept below 50 m/s^2 (row 3 in Fig. 2.2). Novelty in pole length shows a similar qualitative result (row 4 in Fig. 2.2). Most importantly, for these two novelties, agent's performance drops for large values of the respective parameter even if there is no novelty (the diagonal is not even visible in the heatmaps in rows 3 and 4 of Fig. 2.2). Despite the agent's simulation parameters being set to the values of the test world, the agent takes sub-optimal actions and underperforms.

2.5.2   Trained DQN Agent

Next we consider a simple DQN-based agent that runs in the world w for $N = 100$ iterations to train a Deep Neural Network (DNN) to estimate the Q value of each action, and from iteration $N + 1$ and onward chooses actions greedily in the test world $\check{w}$ according to the previously learned network without adapting it to $\check{w}$. Each iteration is an episode or game in which the agent must keep the pole balanced for as long as possible, up to t_{max} time steps, and their performance is computed as t_α/t_{max}, where t_α is the actual amount of time the pole is kept balanced.

Let DNN_i be the trained network at iteration i. The internal state of the agent in iteration $i \leq N$ comprises all the observation-space states collected $(E_{o,t})$ so far and the current DNN (DNN_i). After the N'th iteration, the agent's internal state is only the learned DNN, as the agent chooses its actions according to the DNN and the given observation-space state. Formally,

$$f_t(x : E_t) = \begin{cases} (E_t, DNN_t) & \text{if } t \leq N \\ DNN_N & \text{if } t > N. \end{cases} \tag{2.4}$$

The actions the agent can take at any time t are "left" or "right," which correspond to an instantaneous push of the cart in the corresponding direction with the force F_p.

Results

Next, we evaluate the Deep Q-learning Network (DQN)-based [2] agent that is trained in one world w and tested in another (possibly novel) world $\breve{w}$. As with the simulation-based agent, we vary pole length L, gravity G, push force F_p, and horizontal force F_h to observe the impact on performance as a function of the difference-based dissimilarity between w and $\breve{w}$, $\tilde{\mathcal{D}}^d_{w,\mathcal{T}}(w, \breve{w})$. Figure 2.3 (row 1) shows the results of varying the push force from 0 to 360 N between training (agent's) and testing (world's). The results show that training and testing with significantly different push forces results in increased loss, regardless of whether the difference is an increase or decrease in push force. For the experiments on horizontal force F_h in Fig. 2.3 (row 2), varying horizontal force from -10 to $+10$ N again shows that training and testing on significantly different horizontal forces, especially those in opposite directions, results in increased loss. For gravity G in Fig. 2.3 (row 3), varying gravity from 0 to 360 m/s^2, training in lower gravity results in increased loss; whereas, training in higher gravity results in less loss, demonstrating that the direction of change between the agent's training and the world can be important in determining the outcome. In general, training in more difficult settings, e.g., high gravity, may produce a more resilient agent. This is even more apparent in the regret plot on the right side of row 3 of Fig. 2.3, where the regret is generally below the expected loss and in some cases negative, indicating that the agent trained on high gravity can outperform an agent trained on the correct world gravity. Figure 2.3 (row 4) shows the results of training and testing the adaptive DQN agent on worlds with pole lengths L varying from near 0 to 360 m. Here, the dissimilarity between training and testing has little impact on performance with only the most extreme conditions of testing on very short poles resulting in significant loss, so the agent is nearly perfectly resilient to such novelties.

Overall, the results from the expected loss and regret plots versus dissimilarity show that performance decreases as the world dissimilarity increases, and thus an increase in world-level novelty results in a decrease in the agent's performance. However, this decrease in performance can be one-sided, e.g., training on high gravity results in better performance on low gravity, but the opposite is not true. Also, the variance in the results is high, so these trends are not very strong. Overall, the results show evidence that supports the predictive

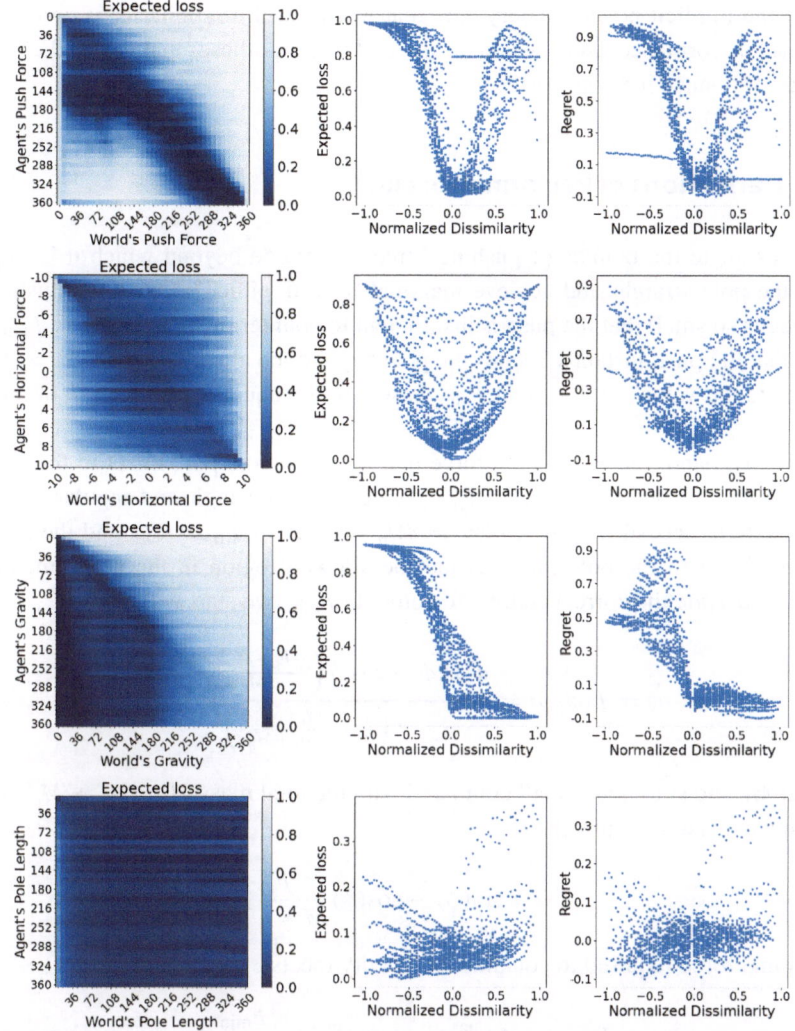

Fig. 2.3 Expected loss and regret of the DQN agent when varying: **a** *push force* from 0 to 360 N, **b** a constant *horizontal force* from −10 to +10 N, **c** *gravity* from 0 to 360 m/s², **d** *pole length* from near 0 to 180 m. The heatmap (left) shows the difference in performance between an agent trained in agent's world, w, and evaluated on a different test world, $\check{w}$. The scatter plots show the expected loss (middle) and regret (right) in the novel environment as a function of difference-based dissimilarity, $\tilde{\mathcal{D}}^d_{\mathrm{w},\mathcal{T}}(w, \check{w})$

power of the implicit novelty theory, i.e., that world-level dissimilarity is correlated with world-level performance, and one could form and test hypotheses making predictions about agents' performance in open worlds.

2.5.3 Derivations of Performance Limits

Here we estimate the bounds of pushing force magnitude beyond which it is impossible to keep the pole straight and achieve maximal reward, either because the pushing force magnitude is so small that the push is insufficient to counter the gravity, $F_p < F_{min}$, or it is so large that the push instantly rotates the pole beyond the allowed region, $F_p > F_{max}$. We approximate the values of F_{min} and F_{max} based on the equations of motion of a CartPole system [1].

On the one hand, the F_{min} is the force that maintains the pole intact in the worst-case initial state, i.e., $x_0^{worst} = (0, 0, \phi_0^{max}, \dot{\phi}_0^{max})$, where both the angle and angular velocity take the most extreme possible values, $\phi_0^{max} = \dot{\phi}_0^{max} = 0.05$. If $F_p < F_{min}$ and the initial state is $x_0 = x_0^{worst}$, then the pole falls beyond the allowed region in the next time step, i.e., $\phi_1 > \phi_0^{max}$, despite the correct action. To compute F_{min}, we start with,

$$\ddot{\phi} = f(\phi, \dot{\phi}) = \frac{g \sin \phi + \cos \phi \left(\frac{-F_p - M_p l \dot{\phi}^2 \sin \phi}{M_c + M_p} \right)}{l \left(\frac{4}{3} - \frac{M_p \cos^2 \phi}{M_c + M_p} \right)}. \tag{2.5}$$

Note that M_p and $\sin(\phi)$ are small compared with the total mass $M = M_p + M_c$, so we can approximate this equation with,

$$\ddot{\phi} \cong g \sin(\phi) - \cos(\phi) \frac{F_p}{M * 1.25}. \tag{2.6}$$

The minimal force required to equalize the gravity meets,

$$0 \cong g \sin(\phi_0^{max}) - \cos(\phi_0^{max}) \frac{F_{min}}{M * 1.25}, \tag{2.7}$$

which yields,

$$F_{min} \cong \frac{g \sin(\phi_0^{max}) M * 1.25}{\sqrt{1 - \sin^2(\phi_0^{max})}} \cong \frac{g \phi_0^{max} M * 1.25}{\sqrt{1 - (\phi_0^{max})^2}} \cong 0.55 * 1.25, \tag{2.8}$$

which corresponds to the performance drop of a simulation-based agent for very small pushing forces (left verge of the heatmap in row one of Fig. 2.2).

On the other hand, the F_{max} is the force that instantly pushes the pole from the nearly worst-case state, $x = (0, 0, 0, 0)$, to the allowed limits of ϕ^{max}. Numerical integrator (semi-implicit Euler method) is governed by,

$$\ddot{\phi}_{t+1} = f(\phi_t), \tag{2.9}$$

$$\dot{\phi}_{t+1} = \dot{\phi}_t + \tau\ddot{\phi}_{t+1}, \tag{2.10}$$

$$\phi_{t+1} = \phi_t + \tau\dot{\phi}_{t+1}. \tag{2.11}$$

Combining the last three equations yields,

$$\phi_{t+1} = \phi_t + \tau\dot{\phi}_t + \tau^2 f(\phi_t, \dot{\phi}_t). \tag{2.12}$$

If the pole is in the steady state, i.e., $x = (0, 0, 0, 0)$, then Eq. 2.5 gives,

$$\ddot{\phi} = f(\phi, \dot{\phi}) \cong \frac{\cos(\phi) F}{M * 1.25}. \tag{2.13}$$

Next, we plug this equation into Eq. 2.12 and seek the pushing force that will move the pole from the steady state to ϕ^{max},

$$\phi^{\mathrm{max}} = \tau^2 \frac{F_{\mathrm{max}}}{1.1 * 1.25}, \tag{2.14}$$

which gives $F_{\mathrm{max}} \cong 720$. We observe that the drop in performance happens for a smaller $F_p = 324$ (right verge of the heatmap in row one of Fig. 2.2). The $F_{\mathrm{max}} \cong 700$ is overestimated, because we did not take into account the limits on z which further restrict the agent and depends on the velocities of pole and cart.

2.6   Conclusions

In this chapter we have applied the novelty theoretic framework to the relatively simple domain of 2D CartPole, in which the agent must keep a pole balanced using just two actions: pushing the cart left or right. Despite the simplicity of the domain, there are numerous novelties that can be defined, and in this chapter we considered four continuous novelties: changes in push force, horizontal force, gravity, and pole length. Results show that the framework is sufficient to represent novelty in this domain and to predict the effects of various kinds and "magnitudes" of novelty on performance due to the dissimilarity between world parameter values.

One may have expected that the larger the difference between the training and test worlds (the "magnitude" of novelty), the lower the performance of agents. Surprisingly, we showed that this expectation is met only for one of the four novelties. Namely, the difference in horizontal force resulted in symmetric performance drops that deepen with the difference between the two worlds for both of the tested agents (row two in Figs. 2.2 and 2.3). It turns out that certain agents are nearly perfectly resilient to other novelties, e.g., the performance of a simulation-based agent is not affected by the change to the push force (row one in Fig. 2.2), whereas the DQN agent is not affected by the change to pole length (row four in Fig. 2.3). We reason that the change in push force does not affect the optimal action, but it influences action effects, so the simulation-based agent is resilient to it, because the agent plans only one step ahead, a change to any action effects does not change the agent's plan. Following this insight, we hypothesize that one-step lookahead agents will be resilient to any novelty meeting this description. In contrast, the non-adaptive DQN agent may be affected by such novelties, because its Q-value function takes into account multiple steps ahead, but its values do not correct for the changes in action effects due to lack of adaptation, so the values may become incorrect, resulting in lower performance (as exemplified in row one of Fig. 2.3). For the same reasons, however, the tables are turned for the two agents when we vary pole length. Namely, stopping a long pole from falling requires a long-term plan and early preventive actions, otherwise it may become impossible to prevent it from falling. Hence, the performance of a simulation-based agent drops for the novelty in pole length (row four in Fig. 2.2), whereas it is nearly unaffected by the DQN agent (row four in Fig. 2.3). Future work can explore adaptive versions of these agents and explain, hypothesize about, and take advantage of the resilience of particular agents to certain novelties to ultimately design agents that respond well to wide sets of novelties.

Another surprising result is that agents experience performance drops for certain parameter values even if there is no novelty, i.e., agents are tested and trained in the same environment. This can happen for two reasons. First, the changes to parameter values may affect an optimal agent's performance and there is no way for an agent to perform better than this, e.g., because the gravity is too high to prevent the pole from falling (observe the heatmap diagonals in row 3 of Figs. 2.2 and 2.3). Thus, when computing regret it is important to take into account the performance of a near-optimal agent or the best available agent. The latter option is more practical, since near-optimal agents are unknown in many domains. Second, the drop in performance due to changes in certain parameters of the environment may reveal inherent limitations of agents, e.g., stopping a long pole from falling requires a long-term plan, which explains why the performance of the simulation-based agents drops drastically for longer pole lengths (the heatmap diagonal in row 4 of Fig. 2.2). We conclude that to understand the impact of novelties on an agent, one must first understand the impact of varying environmental parameter values on that agent's performance and a near-optimal agent's performance, while training and testing in the same environment, i.e., without a novelty.

There were also some surprises that helped to refine the framework. First, the direction of change due to novelty can have an asymmetric effect on agent performance. Further experiments will be needed to explore the properties of domains in which this asymmetry is likely, as well as experiments varying multiple world parameters. Second, we noticed that the observation regret needs access to the world state in order to determine whether the pole will fall or not in the next time step, since the state transition operator acts on the world state and not the observed state. These observations may help in refining future iterations of the novelty framework.

This chapter did not explore the novelty framework in the context of detecting the presence of novelty, nor actively adapting to novelty by updating the agent's internal state in the novel world. Some of these aspects are explored in the next chapter, in which we consider a more challenging 3D version of the CartPole domain.

References

1. Florian RV (2005) Correct equations for the dynamics of the cart-pole system. In: Center for cognitive and neural studies (Coneural), Romania
2. Mnih V, Kavukcuoglu K, Silver D, Rusu AA, Veness J, Bellemare MG, Graves A, Riedmiller M, Fidjeland AK, Ostrovski G et al (2015) Human-level control through deep reinforcement learning. Nature 518(7540):529–533

Novetly in 3D CartPole Domain

3

T. Boult, N. M. Windesheim, S. Zhou, C. Pereyda and L. B. Holder

This chapter expands on the underlying theory and applies it to a 3D CartPole domain. We introduce a Weibull Open World control-agent (WOW-agent) that uses Extreme Value Theory (EVT) to convert dissimilarity to probability of novelty. While novelty is something sufficiently dissimilar to the past experience, that does not mean the WOW-agent cannot reason about the novelty. While there have been many books addressing out-of-distribution detection, novel class discovery, and open-world learning, not all novelty is out-of-distribution data or a novel class. A WOW-agent could be observing known classes taking part in novel actions, known classes involved in novel interactions, or novel changes to known classes, as well as observing novel classes. This chapter expands on the unified model to define more general models of novelty to support multiple simultaneous subtypes, new ways of detecting those novelties, and ways of testing multiple levels of novelty.

3.1 Task Overview

This chapter considers the simulation domain of CartPole3D, allowing us to study different subtypes of novelty. CartPole3D is an extension inspired by the well-known classic control task 2D CartPole [1], described in Chap. 2, where the task is to control the cart using two

T. Boult (✉) · N. M. Windesheim
University of Colorado Colorado Springs, Colorado Springs, CO, USA
e-mail: tboult@vast.uccs.edu

S. Zhou
Cheyenne Mountain High School, Colorado Springs, CO, USA

C. Pereyda · L. B. Holder
Washington State University, Pullman, WA, USA

© The Author(s), under exclusive license to Springer Nature Switzerland AG 2024
T. Boult and W. Scheirer (eds.), *A Unifying Framework for Formal Theories of Novelty*, Synthesis Lectures on Computer Vision,
https://doi.org/10.1007/978-3-031-33054-4_3

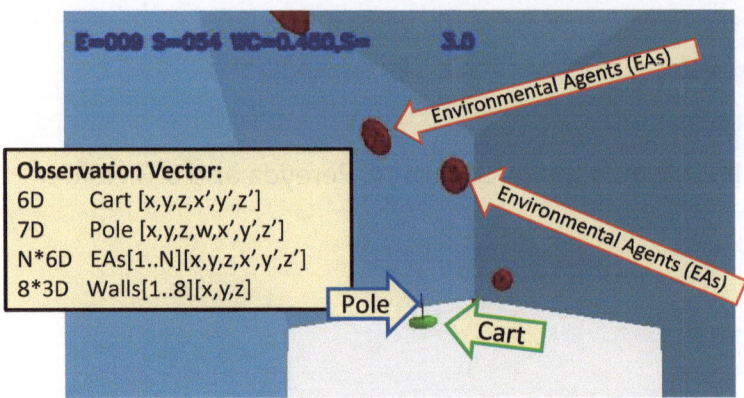

Fig. 3.1 Our multi-type novelty experimental environment uses CartPole3D, where the goal is to keep the pole balanced. The environment has a controllable cart (green) balancing a pole (blue). It also has added independently moving environmental agents (red). The environment can be changed to provide for multiple subtypes of novelty. The Weibull Open World control-agent (WOW-agent) only sees the observational vector of 37–73 numeric values of the position/velocity of the cart, pole, the environmental agents, and the walls that define the world boundaries. In each episode, it receives the observations of each step, makes a control decision (left, right, front, back, none), and tries to keep the pole balanced for 200 timesteps. It reports the probability the world is novel in each episode. The image also shows episode (E=) and step number (S=), the WOW-agent's probability that the world has changed (WC=) to a novel state and the WOW-agent score (S=)

actions, push left and right, to keep the pole balanced for as long as possible or until an episode ends (generally 200 time steps).

The primary difference between the CartPole3D environment and classic CartPole is that CartPole3D lives in a (simulated) three-dimensional world that has flexibility/complexity, and requires the controller to choose between five responses: push front, back, left, right, or no action. While 3D Cartpole environments have been used in other RL-based control studies, [2], our usage is because we add independently moving agents to introduce/study novelty. These are typically depicted as blocks or balls that move around in the environment and can collide with the cart and pole. Different independent agents are introduced into the environment to create new types of novelty (see Fig. 3.1 for an example).

The test environment can introduce novelties where there is a difference in the independent agents, e.g., the number, size, or stickiness of the agent, the actions of the spherical agents, e.g., random motion versus directed attacking motions, or even the interactions (if balls coordinate in action). These would all be novelties the WOW-agent needs to detect and manage. If a WOW-agent can recognize which type of novelty has been introduced, it might impact how they choose to behave, e.g., a sticky/stationary ball could help balance the pole, while attacking independent agents should be avoided. The experiments in this chapter will use this domain, comparing a one-step and two-step lookahead WOW-agent using Extreme

Value Theory (EVT) for multi-type novelty detection with the same WOW-agent using a Gaussian-based novelty detection. We also compare with a Deep-Q-learning Network (DQN)-based control algorithm [3] that is not novelty aware.

3.2   Dissimilarity and Regret

The framework [4] has critical elements defining novelty based on task-dependent dissimilarity functions $\mathcal{D}_{w,\mathcal{T};E_t}$ $\mathcal{D}_{o,\mathcal{T};E_t}$, and associated thresholds δ_w and δ_o for the world and observational space respectively. We generalize this by defining multiple dissimilarity measures in each space using task-semantics to provide multiple types of novelty. First, as shown in Fig. 3.2, we expand the definition to have four dissimilarity operators in each of the world spaces and observed spaces:

- Agent Dissimilarity: $\mathcal{D}_{w,\mathcal{T},A}/\mathcal{D}_{o,\mathcal{T},A}$
- Motion Dissimilarity: $\mathcal{D}_{w,\mathcal{T},M}/\mathcal{D}_{o,\mathcal{T},M}$
- Relations Dissimilarity: $\mathcal{D}_{w,\mathcal{T},R}/\mathcal{D}_{o,\mathcal{T},R}$
- Interaction Dissimilarity $\mathcal{D}_{w,\mathcal{T},I}/\mathcal{D}_{o,\mathcal{T},I}$

In the two-party evaluation used in this chapter, the definitions of these dissimilarity measures are up to the Eval-Team, who defines the worlds with novelties. In the real world, we never know the actual novelty. In this formulation, the Control-Team does NOT get to know how the world/observed dissimilarity measures are defined or even which of the four dissimilarities to use. The Control-Team must detect novelty and estimate novelty subtype based on the combination of its own dissimilarity functions and the roughly implied semantics.

The definitions of novelty can be generalized to as many semantic subclasses of novelty as one wants for the given task. In particular, the control agent defines 64 different dissimilarity dimensions for agent dissimilarity. Probabilities over subsets of these 64 dimensions are combined to approximate the unknown measures for world/observed dissimilarity measures.

Note that this approach is explicitly anticipating particular classes of novelty which we contend is still consistent with the idea of novelty. This approach also uses properties of the observed novelty to determine a more semantically meaningful label. While not explored in this chapter, such a classification may be useful for adapting to that novelty. For example, with finer subtypes/categories such as attacking-motion versus supporting-motion, e.g., one would adapt differently by avoiding the spheres displaying former novel motions and seeking out those with the latter.

The second expansion of the model from Chap. 1 is by adding multiple dissimilarity operators for the agents. While their model provides the observed data to the agent, it is defined by the Eval-Team (our oracle) with their dissimilarity operators in the world space. Hence, we define a dissimilarity operator for agents who can use a mixture of the observed

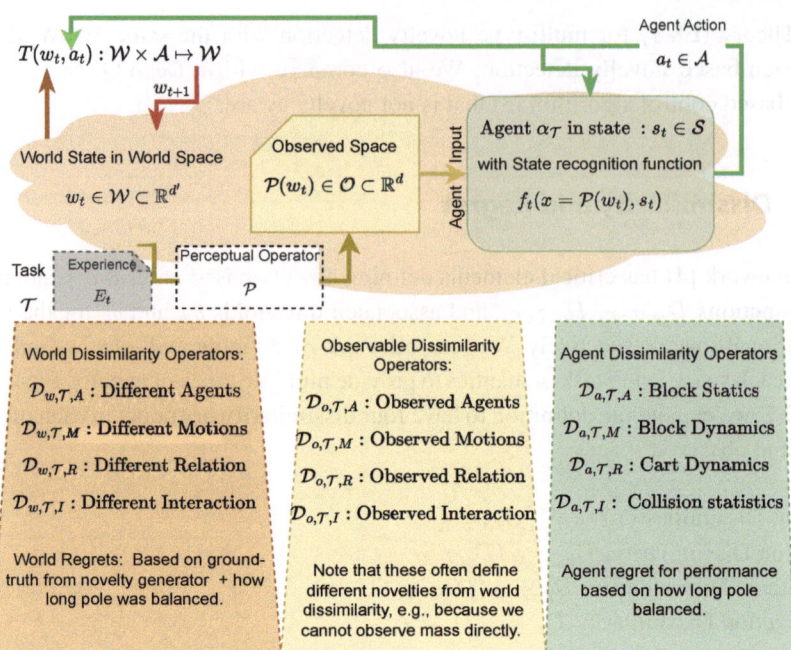

Fig. 3.2 In the framework of Chap. 1 for novelty, the critical elements for defining novelty are task-dependent dissimilarity functions $\mathcal{D}_{w,\mathcal{T};E_t}$, $\mathcal{D}_{o,\mathcal{T};E_t}$, and associated thresholds δ_w and δ_o for the world and observational space respectively. This chapter shows that multiple dissimilarity operators support defining multiple subtypes of novelty, e.g., novel agents, novel actions by agents, novel interrelations, etc. We also define a WOW-agent agent-dissimilarity-operator and show how to use Extreme Value Theory (EVT) to deal with the inherent uncertainty in the world while defining an effective threshold δ_a for the WOW-agent to declare novelty subtypes. Finally, we show that EVT can be effectively used to determine probabilities and the novelty detection thresholds in a 3D CartPole environment

data and the full agent state history. In particular, this allows us to define a formal process for the agent to use an EVT-based process, described in the next section, to estimate a threshold. It is worth noting that the agent dissimilarity may declare something novel where the observed dissimilarity may not. The discrepancy can occur because the agent chooses a limited set of data for computing/normalizing its dissimilarity; in our case, we use a distributional model even if the observed space has a large set of actual states. We also introduce an agent space regret operator. Let $\mathcal{P}_{a,\mathcal{T}} : (x) \mapsto [0, 1]$ be the task performance measure, which for CartPole is the percentage of time the pole was balanced. Then we define agent regret as $\mathcal{R}_{a,\mathcal{T}}(x) = 1 - \mathcal{P}(x)$, i.e., the failure percentage for the task performance operator. While world regret includes novelty, this agent regret has no element for the secondary task of detecting novelty and hence we have only one regret.

3.3   Measurement and Observations

To address the problem of multiple subtypes of novelty, this chapter adopts and extends the framework for defining novelty theory from Chap. 1 (see also [4]) and expands it to formalize the definition of and detection of multiple semantically different levels of novelty. In the original framework, novelty was defined using abstract dissimilarity measures in the world or observation spaces. However, the theory did not define "agents spaces," which are required to build our novelty detection.

The framework in Chap. 1 provides a formal definition of nuisance novelties, which we also adopt as it is useful to be able to distinguish novelties for which there is a novelty but where the world and observation regret differ. For example, in the CartPole3D domain, using just the state interface (versus image interface), the agent can never detect a "cart-color" novelty as it is not observable. Hence, if world regret depends on the detection of that novelty, there will be inherent disagreement.

The core theory in Chap. 1 is a good start, the model is not operational, not precise, and ignored uncertainty/variations. In this chapter we propose an operational definition: a *non-detection nuisance novelty* occurs when the agent regret is not statistically different from its performance in the normal world but when world regret is statistically different. While there may be more general models for statistically significant testing, we use a very simple model. Let $\mu_{\hat{a},\mathcal{R}}$, $\sigma_{\hat{a},\mathcal{R}}$ be the mean and standard deviation in a normal world of the regret of the baseline agent $\hat{a}$. Then an agent is robust if $|\mathcal{R}_{a,\mathcal{T}}(x) - \mu_{\hat{a},\mathcal{R}}| < 2\sigma_{\hat{a},\mathcal{R}}$ where the baseline can be the agent itself but ideally is a near-optimal algorithm so that robustness is not measured with respect to a badly performing algorithm. Let $\mu_{w,\mathcal{R}}$, $\sigma_{w,\mathcal{R}}$ be the mean and standard deviation of normal world regret. Then we declare a novel world instance x to be a nuisance novelty, if and only if, the agent is robust on input x and $|\mathcal{R}_{w,\mathcal{T}}(x) - \mu_{w,\mathcal{R}}| > 2\sigma_{w,\mathcal{R}}$. Intuitively, non-detection nuisance novelty occurs when the world regret is significantly impacted by x while the agent is robust, i.e., the agent regret with respect to task performance is not impacted by the novelty.

3.3.1   Approach to Novelty Detection Using Observations

The problem here has three aspects. First is the core CartPole3D task—keep the pole balanced, the second is the detection of novelty, and the third is the characterization of the subtype of novelty. In this section, we describe the key methods used for each task.

Control. Our WOW-agent is a classic lookahead agent where we consider one-step and two-step lookaheads. To implement a lookahead WOW-agent, we had to augment the CartPole++ 3D world with a set-state function, which was more complex than it might sound. The CartPole++ environment implemented the cart using only a partially implemented (and not officially supported) *planar joint*, which constrains the object to move only within the x-y plane. Unfortunately, the class did not have a sufficient implementation to support even a basic set-state function. The second issue is that in the planar joint model, the cart is defined with a base position plus a joint offset but then returns position via a center-of-mass that is neither of those values. Thus, given the final position, infinitely many combinations can result in the same endpoint. The pole position is specified as attached to the "top" of the cart, but that position is unknown, e.g., because it depends on the unknown size of the cart. Unlike the 2D CartPole environment, where lookahead can have a near-perfect prediction, even in a normal world, the set-state function is inherently noisy, and prediction errors are non-zero and vary by variable. Thus, setting the internal state of the simulator for lookahead is inherently uncertain.

The scoring function is a combination of how far the pole is off vertical plus a novel element added to try to avoid the bouncing balls. Unlike traditional CartPole, we know there will be agents flying around the arena, and in general, if they collide, it will likely impact the pole. Hence, we compute the distance to the closest potential collision point and include an inverse weighted version of that into the score. We note that we cannot avoid collisions with fast agents aiming at the cart because they can move faster than we can get out of the way. Though not included in this book, one way to adapt to novelty may be to adjust the scoring weight for potential collisions to start moving out of the way sooner. Others would be to change the force or angle of the push, but neither was an option in this version of CartPole3D.

3.3.1.1 EVT-Based Probability of Per-Instance Novelty and Overall Novelty Detection

As discussed in the previous section, we use dissimilarity measures + EVT for detection. We broke the dissimilarity measure computation into 34 *static* variables, 9 *dynamic* variables, and 21 *prediction-error variables*. The static variables are the min and max of the absolute value of the initial position of all the objects in the world. To deal with the variable number of agents, we build one static model for agents by combining variables from all agents using the appropriate min and max operators. We compute the min/max of cart/pole/block velocity for dynamic variables. For prediction-error variables, we use the per-dimensional prediction error between the predicted and the actual state after an action, and we do this for the cart and ball positions (3D*2), the cart, pole, and ball velocities (3D*2), and the pole angle (4D). Static dimensionality would include the wall locations if the novel world had different initial wall locations. Since we hypothesized that walls could move as a potential novelty, we include it in the state. Finally, we also keep track of and utilize the collision and failure frequencies.

We run 3,000 trials and collect the dynamics for the first 40 steps of control and then collect the values of all the above variables. We treat each dimension separately and do Weibull fitting using a tail size of 100, producing 64 different Weibull models. At test time, we compute the variables and sum the Weibull Cumulative Distribution Function (CDF) probability from each for the time step. We also track which individual dimensions have significant EVT probability (>0.01) and use that to build our characterization strings discussed below.

While the above gives us a probability of the current time step being "novel," collisions will often produce variable values unlike those seen in training. We limit training to only the first 40 timesteps to reduce the number of collisions. In 3,000 trials, we only see roughly 50 collisions in the training data, which does not provide enough data to characterize the collision effect. Furthermore, since some novelties may reduce impacts, for overall novelty detection, we want something that looks at the distribution of scores independent of the subtype of novelty or variable used and incorporates a model of reliability. Looking at the literature, we decide to build roughly the approach described in [5]. This approach builds a Kullback–Leibler (KL) divergence model, using a truncated Gaussian, from a vector of probability scores. We compute the expected KL divergence over the first 40 steps in training. Then at test time, we do the same and use the probability of exceeding that value as the overall novelty signal that the world has changed. We then accumulate that probability over trials with a blending to smooth out the impact of occasional collisions.

While we have described this use of EVT-based novelty computation in relative CartPole-specific terms, the reader should see how it can be applied to any set of variables in a simulator in both the raw form as well as their prediction error. The task-specific issues are mapping those to the different dissimilarity levels.

WOW Novelty Detection: EVT Per-Step Combined. When using the dissimilarity-based theory, multiple sources of uncertainty impact how to set the novelty detection threshold. The first is the normal uncertainty caused by random variations in the measure. But the second source of uncertainty would be associated with any model assumptions, e.g., if one assumes Gaussian distribution on values that may or may not hold. The important observation of this section is that we remove the second source of uncertainty by using Extreme Value Theory (EVT) because we do not need to assume much about the distribution of dissimilarity in the normal world.

In the [4] framework, they simply declare thresholds for dissimilarity δ_w, δ_o, without providing any meaningful discussion on how they might be defined/estimated. Given the inherent variations and noise in real problems and the undefined scaling issues, defining these to account for uncertainty in the actual task/worlds is critical. In this section, we propose using EVT to define meaningful thresholds, including the agent threshold δ_a.

There are two primary extreme value theorems, and we build from the Fisher–Tippet Theorem, also known as the statistical EVT of the first type. Just as the Central Limit Theorem dictates that the random variables generated from certain stochastic processes follow Gaussian distributions, EVT dictates that given a well-behaved initial distribution of values, e.g., a distribution that is continuous and has an inverse, the distribution of the maximum (minimum) values can assume only limited forms.

Assuming dissimilarity, like distance, is bounded from below by 0, of the three distributions only the Reversed-Weibull can apply for maximum (and hence Weibull for minima). This EVT, and in particular the Weibull-based fitting, is widely used in many fields [6], such as manufacturing, e.g., estimating time to failure, natural sciences, e.g., estimating 100- or 500-year flood levels, and finance, e.g., portfolio risks. EVT has recently been (re)introduced and applied in recognition, machine learning, and computer vision [5, 7–9] where it is often used in open-set or open-world recognition tasks.

We note that some of the fields for CartPole modeling are not bounded a priori, e.g., maximum velocity is not easily bounded. Such fields might be better modeled with a different type-1 EVT distribution, e.g., Fréchet or Gumbel, or may even be better modeled with a peak-over-threshold, i.e., EVT type 2 [10, 11], approaches. However, for simplicity we have used Weibull modeling for all fields and found it to be sufficiently effective and leave the search for better per-field models to future work.

To apply the Weibull to determine a novelty detection threshold we consider some dissimilarity measures and collect values of this measure of a large number of trials in the normal world. We can then use the largest or smallest of these scores, depending on the variable type, and fit a three-parameter Weibull distribution to them,

$$
W(x; \mu, \sigma, \xi) = \begin{cases} \frac{\xi}{\sigma} \left(\frac{x-\mu}{\sigma} \right)^{\xi-1} e^{-\left(\frac{x-\mu}{\sigma} \right)^{\xi}} & x < \mu - \frac{\sigma}{\xi} \\ 0 & x \leq \mu, \end{cases}
$$

where $\mu \in \mathbb{R}$, $\sigma \in \mathbb{R}^+$, and $\xi \in \mathbb{R}^-$ are locations, scale, and shape parameters. Multiple libraries can compute the three-parameter Weibull, including SciPy, used for this book. The associated CDF is given by,

$$
Wcdf(x; \mu, \sigma, \xi) = \begin{cases} 1 - e^{-\left(\frac{x-\mu}{\sigma} \right)^{\xi}} & , x < \mu - \frac{\sigma}{\xi} \\ 0 & , x \leq \mu, \end{cases}
$$

which can be used to compute the probability of novelty for any given dissimilarity score x. In our control system, we accumulate this probability over different items and then threshold ($p = 0.99$), which simplifies processing varying numbers of dissimilarity. If desired, one can use the distributional parameters μ, σ, ξ, to compute the dissimilarity threshold δ that yields a given probability $0 < p < 1$ such that probability $(\mathcal{D}_{a,\mathcal{T}} > \delta) \geq p$ implies novelty. We use the inverse CDF to derive $\delta_a = \mu + \sigma * (-ln(1-p))^{\frac{1}{\xi}}$.

KL-Divergence Over Per-Episode Novelty-Detection. The per-dimension approach produces a vector with per-time step probabilities of novelty. If the simulation and predictions were perfect, thresholding such per-time step data may be sufficient to detect novelty. However, we found probability from a single trial was too noisy to use directly. This is because the 200 time steps per episode and 100 s of episodes per trial, combined with the noise inherent in the simulation with a imperfect set-state function, have too high a probability of false detection. To improve robustness, we consider the distributions of the novelty probabilities over time and inspired by [5]. We use KL-divergence to compare the normal distribution to the distribution from the test data (see Fig. 3.3). To improve sensitivity for small differences, we also use the accumulation of novelty probability over time.

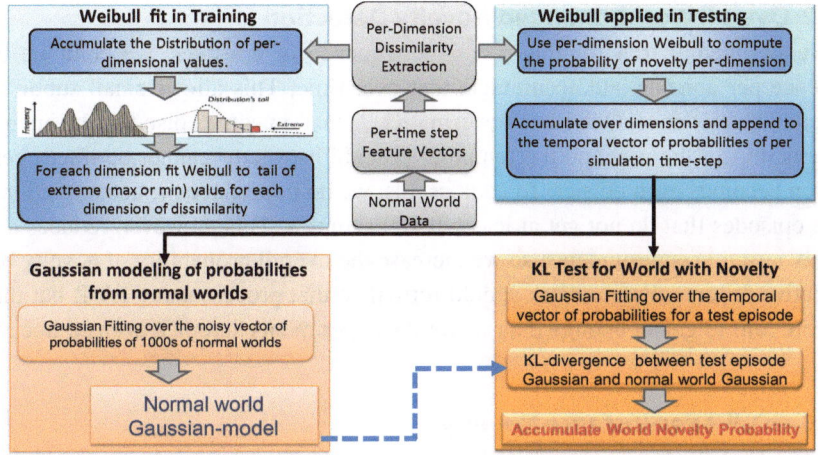

Fig. 3.3 Overall novelty processing per instance and per episode. Data is collected from each dissimilarity dimension during testing, and Weibulls are fit to the extreme value for each. Given a trial, at each time step, the dissimilarity values are computed and the per-dimensional Weibull models yield probabilities that accumulated over dimensions and appended to a temporal probability vector. For 300 normal trials, i.e., examples of normal worlds, the system builds a Gaussian model (mean/standard deviation) of the elements of the temporal probability vector. For each episode of a test trial, the temporal probability vector is computed and KL-divergence is computed comparing that to the Gaussian of the normal world. The per-episode probabilities are accumulated over time to allow for high sensitivity with low false detection rates

The Kullback–Leibler (KL) divergence is a fundamental measure of the difference between distributions. It measures the relative entropy,

$$KL\,(P\|Q) = \int_{-\infty}^{+\infty} p(x)\,\log\left(\frac{p(x)}{q(x)}\right)\,dx, \tag{3.1}$$

where $p(x)$ is the probability density function of the testing vector, and $q(x)$ is the probability density function of normal world vectors. In our case, both the training and test vectors are distributions of per-time novelty probabilities and the goal is to detect changes in that distribution. Intuitively, the training vector would be all small values and the test all large values, but KL allows us to formalize "how" different are the distributions.

Making the classic Gaussian assumption for the pair of distributions, i.e., letting $p(x) \sim \mathcal{N}(\mu_t,\ \sigma_t^2)$ and $q(x) \sim \mathcal{N}(\mu_n,\ \sigma_n^2)$, be the distribution parameters of the test and normal worlds respectively, the KL divergence measure is,

$$KL\,(P\|Q) = \log(\frac{\sigma_n}{\sigma_t}) + \frac{\sigma_t^2 + (\mu_t - \mu_n)^2}{2\,\sigma_n^2} - \frac{1}{2}. \tag{3.2}$$

3.3.1.2 Overall World-Changed Novelty Detection

As an experimental trial for this task is a sequence of episodes, we compute the KL-divergence per episode and accumulate that over time. This allows small subtle changes to accumulate evidence. One issue for using KL is that, in some episodes, the environmental agents impact the cart and the temporal probability vector cannot be filled; hence KL detection becomes even noisier. To address this, we do not reduce the accumulation weight for KL episodes that do not get at least 40/200 samples. This, however, reduces detection in highly unstable environments, so we increase the overall probability of novelty based on consecutive failures. Future work should formalize that process using EVT, but that takes way more runs to gather enough data, so for the experiments herein, it is an ad-hoc increase.

3.3.1.3 GOWN Per-Instance Novelty

The above described is the core WOW system. For novelty detection, we also developed a Gaussian-based Open-World Novelty (GOWN) baseline. It is identical to the EVT agent in all respects except that the per-instance novelty is based on truncating a Gaussian model. We computed the mean μ and standard deviation σ on each, similar to the EVT and used them in the Gaussian CDF,

$$\Phi(x; \mu, \sigma) = \frac{1}{\sqrt{2\pi}} \int_{-\infty}^{x} e^{\frac{-(t-\mu)^2}{2\sigma^2}} \, dt. \tag{3.3}$$

We could compute probability of an outlier for each dimension as $P_{G1}(x) = 0.5 - \Phi(x; \mu, \sigma)$ and then could use either L^1-norm or L^∞-norm, i.e., sum or max, to combine the dimensions. Unfortunately, both resulted in 100% of the test trials having false positives. Since trials have 200 episodes, with 20–40 episodes before novelty is introduced, noise makes an outlier too likely to occur randomly. To reduce the impact of the accumulation of many small- to medium-sized outliers, we modified the approach to use,

$$P_G(x) = \begin{cases} 0.5 - \Phi(x; \mu, \sigma) & |x - \mu| \geq 3\sigma \\ 0 & \text{Otherwise,} \end{cases} \tag{3.4}$$

which truncates the Gaussian outlier probability when the input is within 3σ of the mean. If the data was Gaussian, this would be an effective threshold of about 0.3% of the data. We fed these probabilities into the KL-divergence detection algorithm described next with its expected mean/variance for overall change detection.

3.3.1.4 Characterization of Per-Instance Novelty

Using the per-dimension EVT-based probabilities, we can characterize the novelty into subtypes. We can combine the static variables to form our agent dissimilarity measure while combining the "velocity" dimensions into a motion dissimilarity measure. We use the remaining dynamic variables for relation dissimilarity, which we expect to be largely driven by prediction error. Hence, we should capture changes in relationships such as sloping of the cart, adding wind or friction, etc. The control agent outputs a string when it reports the world has changed to be novel. It summarizes the detected deviations along the lines of "Initial world off and Dominated by Balls with 18 Velocity Violations 18; 18 Agent Velocity Violations; 30 Total Agent Violations; 0 Cart Total Violations; 0 Pole Total Violations ..." which would be overall a motion violation (agent velocities).

3.4   Novelty Types and Examples

The WOW-agent developed for this research seeks to detect previously unseen novelties and classify them into different semantic novelty subtypes. In particular, we consider novelty in classes, agents, actions, relations, and interactions (see Fig. 3.2). Thus, we extend the framework of Chap. 1 [4] to include agent novelty, with a computable EVT-based dissimilarity in the agent space. Secondly, we extend it to multi-type novelty by having multiple dissimilarity measures in all three spaces defined in that chapter: *world space, observed space, and agent space*. These are logically defined in the associated spaces and allow us to have different dissimilarity measures defining the different subtypes of novelty. Our WOW-agent can classify previously unseen novelties using the multiple dissimilarity measures and using that classification to formulate responses.

The observation spaces and agent spaces include random variations, and dissimilarity measure computations will have noise. Hence, there is uncertainty in deciding if this dissimilarity is from noise or is caused by a novelty. A robust detection process is required. We show how to use EVT to determine when a measured dissimilarity is sufficiently different from training to declare something novel. Using the same dissimilarity variables/structures as our EVT-based solution, we also detect novelty using Gaussian-based uncertainty models for comparison.

3.5   Experiments

In multi-agent problems, the term agent can become overloaded; we use the term WOW-agent to refer to our software agent that controls the cart while detecting/managing novelty. We also have a DQN-based agent and a Gaussian-control agent, which are baseline software agents that control the cart and detect novelty, respectively. We use the term environmental agents to refer to other independently moving agents in the environment.

We also have two teams involved in the research: we refer to the Eval-Team as the team that generated the core environment and environmental agents, as well as, ran the sequestered tests. We use Control-Team to designate the team that created the controlling agent that needs to control the cart while also detecting novelties in the open-world.

Our novel CartPole3D environment has been constructed to utilize BulletPhysics [12], with a mixed C++/Python code base. The Eval-Team and Control-Team developed a slightly modified version of pyBullet that uses a modified planar constrained joint for the cart and to support setting the state of the position and joints used for the cart. The latter was a critical function for the lookahead controller used by the Control-Team. This was incorporated into a python environment used by both teams – in particular, this chapter uses v0.7.9 of the system from https://github.com/holderlb/WSU-SAILON-NG. The resulting CartPole3D model is a simulated environment with roughly 200 total changeable parameters/properties/attributes, some of which have easily changeable python interfaces, others are in model "urdf" files, and more are changeable only in the C++ code.

Figure 3.1 shows a graphic rendering of the basic CartPole3D environment, the cart, the pole, the walls, and an example with four environmental agents. While the simulator can generate images, in this project we do not receive images, just low-dimensional observation vectors, which is all the information the controlling agents receive about the environment.

For the cart controllers, we consider three different models:

- DQN-agent: Deep-Q-learning Network (DQN)-based control algorithm, which is the baseline
- GOWN-agent: Gaussian-based Open-World Novelty agent
- WOW-agent: our Weibull-based Open-World control agent

The Eval-Team developed the DQN-agent for CartPole3D to serve as a baseline. The baseline DQN-agent was a standard deep neural network with two layers of 512 fully connected nodes with rectified linear unit activation. The agent used DQN training, similar to that in [3, 13, 14]. The agent was trained on non-novel CartPole3D episodes and converged after 500K steps. The final average performance reached was 0.8 out of 1.0, which represents keeping the pole balanced for 160 time steps on average out of the possible 200 time steps.

In our experiment, we used a sequestered evaluation methodology so that the Cart-Pole agent subteam does not know the actual novelties. The book's authors represent two non-overlapping groups at different organizations. The evaluation subgroup designed/ implemented the novelties and evaluated the systems, while the control agent subgroup developed the CartPole3D novelty detection agent. Our long-term goal is an automated evaluation system and baseline agent for comparison.

The experiments had 360 runs with four subtypes of novelty. In addition, the Eval-Team defined three difficulty levels per subtype, by varying the world dissimilarity of the novelties they introduced, with smaller dissimilarity being harder to detect. Overall, there were 30 random sample runs per "subtype/difficulty" setting. Each run had 200 trials for 360*200 total trials. Each trial consisted of 200 CartPole episodes where a random K (unknown to control agent) number of normal episodes was followed by 200-K novel episodes. For each episode t, the performance scores $S_t(A)$ was the fraction of the 200 time steps where agent

A kept the pole within the required range, the run terminated with performance =1, if the agent successfully reached 200 steps.

To evaluate the system we considered the following metrics, computed over the 30*200 trials per the subtype/difficulty,

FP Rate: Fraction of trials where a false positive (or early detection) was reported. It was always zero for EVT experiments but not for Gaussian and so reported for completeness.

Detection rate: Mean± stdev fraction of correctly detected trials. Note that a trial is considered a false positive and not a correct detection if the agent declared novelty before novelty was introduced.

Detection Delay: Mean± stdev number of episodes after novelty was introduced before it was declared detected.

Performance: Mean± stdev of performance score $S_t(C)$.

OPTI: (Overall Performance Task Improvement) Mean± stdev of relative post-novelty performance to a baseline DQN WOW-agent,

$$OPTI = \frac{\sum_{t \in \text{novel}} S_t(C)}{\sum_{t \in \text{novel}} S_t(C) + \sum_{t \in \text{novel}} S_t(B)},$$

     where C is the CartPole control-agent being tested (EVT or Gaussian) and B is the baseline DQN-agent.

Robust%   Which is the percentage of trials that were not statistically different from normal, using agent performance in a normal world.

The scoring of characterization was done by the Eval-Team by matching the reported strings with the ground-truth novelty introduced for that trial. Human interpreters scored the accuracy of the objects involved, e.g., balls, cart, etc., the property or relationship changed, e.g., movement, and the change type, e.g., increased speed or direction change. In the performance table, we list the characterization of object accuracy and characterization of property accuracy.

The Control-Team implemented the WOW-agent and GOWN-agent; see the MDPI branch of git@github.com:Vastlab/SAILON-CartPole3D.git. Both used a lookahead-based control algorithm, where it received the current observational vector and compared that with its projected state. We noted that to support the Control-Team state setting functions needed for the WOW-agent lookahead type controller, the Control-Team developed a modified pyBullet. The state setting was imperfect as the test system only provided truncated position and velocity of the cart, pole, and active agents—which produced insufficient information to exactly recreate the state. Thus, the lookahead state prediction was inherently noisy, and the WOW-agent must detect novelty in the presence of a noisy world model.

The GOWN novelty detector was weak, with an average FP rate over 30% on all subtypes of novelty. Its detection rate was much lower on agent and action subtypes of novelty. Using

Table 3.1 Performance of EVT-based detection with one- and two-step lookahead WOW-agent. FP rate was 0 for all EVT-based tests. The detection performance was perfect for agents and actions and near zero for relations and interactions – the latter were, however, nuisance novelties as robustness was near 100%. The performance of the DQN WOW-agent on the same trials is shown at the far right of table. The difference is statistically significant

Novelty subtype	Difficulty	Detection rate	Detection delay	OPTI	Object	Property	Robust percent	WOW agent performance	DQN performance
Agent	Easy	1.0	3.77	0.56	1.0	1.0	0.98	0.83	0.67
	Medium	1.0	3.73	0.56	1.0	1.0	0.92	0.75	0.60
	Hard	1.0	3.77	0.54	1.0	1.0	0.88	0.64	0.53
	Subtotal	1.0	3.76	0.55	1.0	1.0	0.93	0.60	0.74
Action	Easy	1.0	15.27	0.51	1.0	1.0	0.80	0.63	0.60
	Medium	1.0	6.40	0.48	1.0	1.0	0.42	0.44	0.47
	Hard	1.0	4.03	0.47	1.0	1.0	0.17	0.32	0.37
	Subtotal	1.0	8.57	0.49	1.0	1.0	0.46	0.48	0.47
Relation	Easy	0.0	0.0	0.54	0.0	0.0	0.98	0.89	0.75
	Medium	0.0	0.0	0.55	0.0	0.0	1.00	0.88	0.72
	Hard	0.10	7.43	0.55	0.0	0.0	1.00	0.85	0.70
	Subtotal	0.03	2.48	0.55	0.0	0.0	0.99	0.72	0.87
Interaction	Easy	0.0	0.0	0.53	0.0	0.0	0.98	0.89	0.79
	Medium	0.03	5.10	0.54	0.0	0.0	0.98	0.90	0.78
	Hard	0.0	0.0	0.54	0.0	0.0	0.98	0.90	0.78
	subtotal	0.01	1.70	0.54	0.0	0.0	0.98	0.90	0.78

a two-sided paired t-test, the GOWN detector was significantly worse than the EVT detector ($p < 0.01$) in both FP rate and detection rate. We did not score the Gaussian detector for characterization, but given the much worse detection, we expect characterization will also be much worse.

3.6 Conclusions

This chapter introduced an example extension to the main framework to support multiple subtypes of novelty, using different dissimilarity measures for each. The core idea could be applied in almost any problem with appropriate adaption of the dissimilarity measures. The chapter also showed how to use Extreme Value Theory to detect when the different dissimilarity measures are sufficiently different from the "normal" training data. WOW-agent using EVT was effective in detecting some of the novelty subtypes and not effective in detecting others. The latter are examples of nuisance novelty, as shown by the "robustness" column in Table 3.1, poor detection corresponded to high robustness. However, for agent novelty, the detection was still high despite the robustness suggesting that when the dimensions used in dissimilarity contain sufficient information the system can still detect "nusiance" novelties.

Fundamental to the unified framework, is the use of to dissimilarity measures to define novelty in different spaces. The EVT-based models presented in this chapter, and the next, provide a robust statistically grounded approach to dealing with the difficult problem of determining when the dissimilarity is sufficiently large. We believe EVT provides a basis for almost any novelty detection problem in the unified framework. While this chapter has focused on the use of Weibull models, it is worth nothing that there are other EVT distributions and theories, which is most appropriate depends on the task and dissimilarity measures. However, in the unified framework, novelty can always be cast as some type of "extreme" dissimilarity.

References

1. Brockman G, Cheung V, Pettersson L, Schneider J, Schulman J, Tang J, Zaremba W (2016) Openai gym
2. Xiao T, Jang E, Kalashnikov D, Levine S, Ibarz J, Hausman K, Herzog A (2020) Thinking while moving: deep reinforcement learning with concurrent control. In: International conferences on learning representations (ICLR). arXiv:2004.06089
3. Lillicrap TP, Hunt JJ, Pritzel A, Heess N, Erez T, Tassa Y, Silver D, Wierstra D (2015) Continuous control with deep reinforcement learning. arXiv:1509.02971
4. Boult TE, Grabowicz PA, Prijatelj DS, Stern R, Holder L, Alspector J, Jafarzadeh M, Ahmad T, Dhamija AR, Li C et al (2021) Towards a unifying framework for formal theories of novelty. In: Proceedings of the AAAI conference on artificial intelligence, vol 35:17, pp 15047–15052
5. Jafarzadeh M, Ahmad T, Dhamija AR, Li C, Cruz S, Boult TE (2021) Automatic open-world reliability assessment. In: Proceedings of the IEEE/CVF winter conference on applications of computer vision (WACV), pp 1984–1993
6. Kotz S, Nadarajah S (2001) Extreme value distributions: theory and applications. World Scientific Publishing Co
7. Scheirer WJ (2017) Extreme value theory-based methods for visual recognition. Synth Lect Comput Vis 7(1):1–131
8. Gibert X, Patel VM, Chellappa R (2015) Sequential score adaptation with extreme value theory for robust railway track inspection. In: Proceedings of the IEEE international conference on computer vision workshops, pp 42–49
9. Carpentier A, Valko M (2014) Extreme bandits. In: NIPS, pp 1089–1097
10. Leadbetter MR (1991) On a basis for peaks over threshold modeling. Stat & Probab Lett 12(4):357–362
11. Smith RL (1984) Threshold methods for sample extremes. In: Statistical extremes and applications. Springer, pp 621–638
12. Coumans E, Bai Y (2016–2021) Pybullet, a python module for physics simulation for games, robotics and machine learning. http://pybullet.org
13. Mnih V, Kavukcuoglu K, Silver D, Rusu AA, Veness J, Bellemare MG, Graves A, Riedmiller M, Fidjeland AK, Ostrovski G et al (2015) Human-level control through deep reinforcement learning. Nature 518(7540):529–533
14. Kumar S (2020) Balancing a cartpole system with reinforcement learning–a tutorial. arXiv:2006.04938

Novelty in Image Classification

4

A. Shrivastava, P. Kumar, Anubhav, C. Vondrick, W. Scheirer, D. S. Prijatelj,
M. Jafarzadeh, T. Ahmad, S. Cruz, R. Rabinowitz, A. Al Shami and T. Boult

In this chapter, we introduce real-world data in the form of RGB images and expand on the application of the underlying theory. We divide the world of images into known and novel sets and then use a sampling process to generate a large number of novelty experiments. There is a learning-based novelty-aware classification agent, but it does not actively interact with the world. For the image classification task, the perceptual operators are defined as the features computed from a Deep Neural Network (DNN) trained on the classification task using the known classes.

4.1 Task Overview

In computer vision, the open-world recognition problem [1] is to determine a class label for any input or label it with a label of *novel* while learning to label novel items from the same class with the same label. There are many image recognition tasks that can be per-

A. Shrivastava (✉) · P. Kumar · Anubhav
University of Maryland, College Park, MD, USA
e-mail: abhinav@cs.umd.edu

C. Vondrick
Columbia University, New York, NY, USA

W. Scheirer · D. S. Prijatelj
University of Notre Dame, Notre Dame, IN, USA

M. Jafarzadeh · T. Ahmad · S. Cruz · R. Rabinowitz · A. Al Shami · T. Boult
University of Colorado Colorado Springs, Colorado Springs, CO, USA

© The Author(s), under exclusive license to Springer Nature Switzerland AG 2024   37
T. Boult and W. Scheirer (eds.), *A Unifying Framework for Formal Theories
of Novelty*, Synthesis Lectures on Computer Vision,
https://doi.org/10.1007/978-3-031-33054-4_4

formed within this regime including classification, detection/localization, counting, etc. In this chapter, we focus on classification as the primary task, including learning representations of the base classes from samples of labeled examples. Because only one label is supplied and many items exist in the image, the world dissimilarity may not be consistent with an overall human interpretation, which is why a top-N measure is often used in scoring the task.

In the current setting, the task has 413 known classes (K) and an unknown number of items from an unknown number of novel classes. While the primary task is image classification, there are also task variations. Simple variations are open-set novelty detection, which requires returning a label as one of 414 total classes (the 413 known items + an unknown class), and open-world classification, which must return one of the $413 + N$ labels where there are M novel classes (and ideally $N = M$). We will use the former form of the task for this experiment, which we call the $(K + 1)$-way classification.

There is also an auxiliary "change detection" task where the goal is to detect when, in a sequence, the distribution changes from the training distribution to one with the novelties introduced. To reduce the impact of noise in per instance-level detection, the latter task may use windows of data to estimate distributional shifts better, e.g., see [2].

4.2   Dissimilarity and Regret

- **World $\mathcal{W}$ and the world dimensionality d'**

    The world is the subset of the objects in the physical world, as sampled in terms of selected classes and embedded in images. It can be viewed as a set of class labels + attributes at each point in 3D space, and within a volume, one gets a tensor of such labels. However, in our problem, there is a single label applied to the volume of data, with the label provided by a human labeler. The subset $\mathcal{W}$ and its dimensionality are defined exclusively by the training examples combined with the human annotator's labeling/grouping of those examples.

    To design a novelty experiment, one can separate the world by class labels and withhold data for the "novel items" based on hand-labeled data, computer-generated labels, or both. As the task is classification, the world is varied in dimensions that might not impact human labeling. Thus, the introduced novelties might be ignored or cause humans to label them differently, e.g., a cartoon dog might not be grouped with dogs but instead treated as a different category. However, since people are consistent in their annotations, the task is well-defined and comes to resemble the same labeling decisions humans use.

- **Family of Perceptual Operators $\mathcal{P}_t$ for Each Time Step and Observation Space $\mathcal{O}$ Accessible to the Agent With Dimensionality d**

    The perceptual operators $\mathcal{P}_t$ simply image the world at that time. There is no temporal coherence and the imaging transformation can be, and generally are, different for every sample.

The observed space is the input images of $M \times N \times 3$. It is subject to general imaging conditions, including lighting, occlusion, blur, etc. One of the major issues with sampling for image classification is that the samples contain many different objects/attributes and hence don't have a unique label as presumed for the task. Humans label based on the observed space but assume that a single label applies to the world space.

The agent converts the input image into its internal representation, such as a projection into a deep feature vector. The internal representation can be incrementally updated as new data is collected. From an operational standpoint, in this paper the agent's features are computed with a DNN trained on the classification task using the known classes. The feature space may be kept constant or may be adapted as new examples are seen.

- **Task-dependent World Dissimilarity Functions $\mathcal{D}_{w,\mathcal{T}}$ and Associated Threshold δ_w**

  The dissimilarity function for individual samples of the world model is based on agreement with ground-truth labels. This is implemented as a binary match/no-match with the available ground-truth.

  While the experiments herein use dissimilarity measures considering single samples and novel classes, the framework is general enough for other novelty forms. Given a set of samples, experiments could change the sampling strategy and frequency of selecting items from the set, which is tantamount to a distributional shift in the world, i.e., a distributional novelty. This can be extremely subtle to detect because the variation within the sample set can be large compared to the change in distribution. No formal dissimilarity measure for a distribution has been implemented, but something like total variation, Wasserstein distance, or Kolmogorov–Smirnov distance to estimate distributions from the discrete data samples.

- **Task-dependent Feature Space Dissimilarity Functions $\mathcal{D}_{\mathcal{O},\mathcal{T}}$ and Threshold $\delta_{\mathcal{O}}$**

  This can be the same dissimilarity measure as the world space using ground-truth labels. However, because of potential issues of learning from limited data and multiple potential labels (see Fig. 4.1), a more predictive dissimilarity measure can be obtained via learning by considering a deep network trained on the data in the experience tensor. If we let $f_i(w) \in [0, 1]$ be the predicted confidence the input is from class i then we can define a dissimilarity $\mathcal{D}_{\mathcal{O},\mathcal{T}} = 1 - \max_i f_i(w)$. For the example results below, dissimilarity comes from a ResNet-50 [3] trained for $413 +$ background classes, starting with an experience tensor with 481,416 images for the 413 known classes and 5,000 known unknown images for the background class.

- **Family of State Recognition Functions $g_{y,t}(x = \mathcal{P}(w) : E_t)$**

  A state recognition function could be a deep network classifier with a reject option, which rejects input that are novel. In the examples below, we use the EVM [4], which uses Extreme Value Theory-based Weibull probability measures to a set of stored representative points (extreme vectors). These can be updated as new instances are learned.

- **Family of Actions $a_t(x)$ Taken by Agent A_g**

  This task permits no actions except returning a prediction for the task, i.e., a class label

Cherry	Agamma	Person
Attr: Pink Eyes	Attr: Open Door	Attr: Green Tires

Fig. 4.1 Examples showing the problem of using image labels as definitions of novelty. The official ImageNet labels are under each image, along with example attributes. The potential of many world labels, but only one assigned label used for dissimilarity, can lead to a type of nuisance novelty if people/systems often disagree on the best label—the regret measure will be flawed. This could be partially ameliorated with a type of top-K match criterion, e.g., top-5 is common for ImageNet since the images contain multiple objects. When the sample is considered "novel" in terms of "regret" because of "non-class" properties such as attributes (pink-eyes, green-tires) or state (open-door), which are not directly related to classification and the dissimilarity measure (based on how images were originally labeled), there will also likely be nuisance novelties

as one of the 413 known items or a label indicating novelty for open-set novelty detection, or one of the $413 + K$ labels for the open-world classification task, or the prediction of when the distribution changed.

- **World Regret Function** $\mathcal{R}_{w,\mathcal{T}}$
  This is an error measured in the world space of ground-truth labels and is not directly accessible to the agent. For per instance novelty detection we might consider the actual label versus the reported label. However, we lack any real data about the actual world, so we don't consider world regret.

- **Observation Space Regret Function** $\mathcal{R}_{\mathcal{O},\mathcal{T}}$
  In terms of the experiment below, "measurements" depend on the task and the images. Because of the inherent difficulty of the problem, where even the best classifiers cannot correctly label every instance, we consider relative performance. For the open-set task we consider the relative performance of the algorithm after novelty is introduced with the baseline system performance before novelty, $M_{NRP} = \dfrac{\text{Acc(post,EVM)}}{\text{Acc(pre,}\beta\text{)}}$, where β is the baseline network. Let regret be $\mathcal{R}_{w,\mathcal{T}_\infty} = 1 - M_{NRP}$. For the change detection task, we consider the percentage of correctly detected trials %CDT, and for regret $\mathcal{R}_{w,\mathcal{T}_\in} = 1 - \%\text{CDT}$.

4.3   Novelty Types and Examples

We have three types of novelties, all of which are *world* novelties and are discussed below.

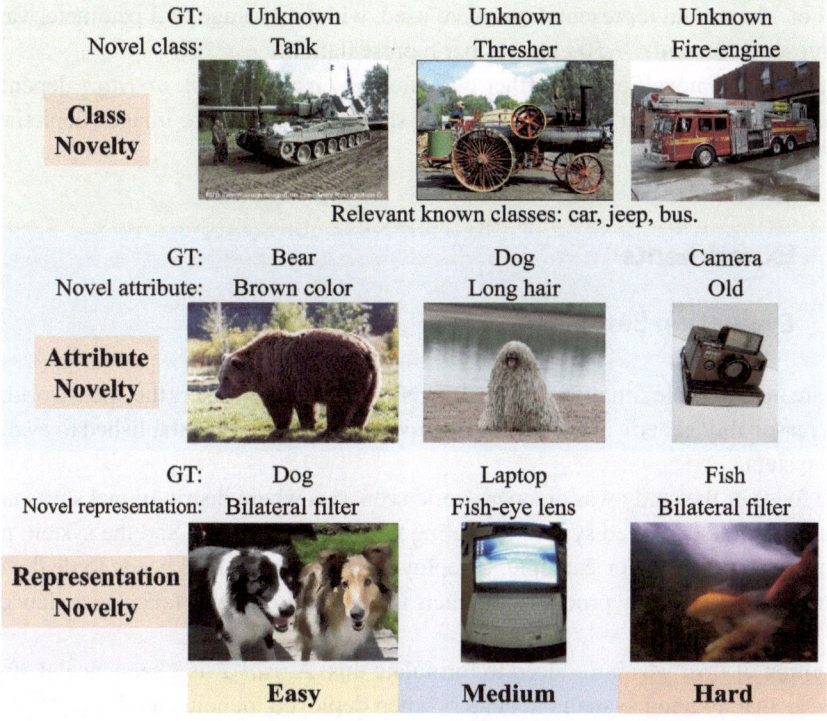

Fig. 4.2 Examples of different kind of novelties at different difficulty levels

Class Novelty. This novelty generates images from unseen classes. We used 160 novel classes with ~1,500 images per class and classes with varying similarities to the 413 training classes.

Attribute Novelty. This novelty evaluates systems' generalization capabilities for known classes by introducing images of known classes composed with previously unseen attributes [5, 6]. For example, the training data for 'dog' does not have any dogs with 'long legs,' whereas D_{post} contains images from the Greyhound breed with 'long legs,' and the system is supposed to classify them as the known class: 'dog.' We used novel attributes for 150 known classes with 300 attribute variants, an average of 3 variations per known class, and ~1,500 images per variation. This was accomplished using a combination of held-out sub-classes for the parent known class and contemporary image-to-image translation generative models [7–9].

Representation Novelty. This novelty captured unseen variations external to the semantics of the object. These included a variety of JPEG compressed images (often a failure mode of contemporary models [10]), climate conditions, e.g., rain, snow, dusty; often encountered by deployed systems, filtered images, e.g., bilateral filters which remove texture that contemporary models are sensitive to [11, 12], lens distortions, e.g., fish-eye [13], motion blur, etc.

A total of 12 external representations were used, with an average of 3 parameter variations per representation and ~7,500 images per representation.

Each of these novelties are further classified as *Easy, Medium, or Hard* depending on various factors (See Sect. 4.4.1). Figure 4.2 shows representative images depicting these novelty types.

4.4 Experiments

4.4.1 Evaluation Setup

There are innumerable situations in which a vision system encounters the open-world. Therefore, a reasonably generic and tractable open-world scenario was established to evaluate the vision system.

The focus of the study was on open-world scenarios where distributional shifts naturally happen while the deployed system is dealing with streaming data, and the system needs to respond appropriately. For example, a deployed product classifier needs to deal with new versions of products, new products, products in new environments (such as review photos), etc.

A notion of *trial* was introduced to formulate this. A *trial* $\mathcal{T}$ is a sequence of streaming images x_t that a vision system encounters when deployed, denoted as $\mathcal{T} : \{x_t\}_{t=1}^T$. During a trial, a singular event of open-world distribution shift can occur at any time step t_{OW}, such that $\{x_t\}_{t < t_{\text{OW}}} \sim \mathcal{D}_{\text{pre}}$ and $\{x_t\}_{t \geq t_{\text{OW}}} \sim \mathcal{D}_{\text{post}}$, where $\mathcal{D}_{\text{pre}}$ is the closed-world train/test distribution as in the standard machine learning paradigm, and $\mathcal{D}_{\text{post}}$ is post-shift distribution. Note that for most interesting distribution shift scenarios, $\mathcal{D}_{\text{post}} \setminus \mathcal{D}_{\text{pre}} \neq \phi$, i.e., the open-world distribution has overlap with $\mathcal{D}_{\text{pre}}$ (Fig. 4.3).

The performance of a vision system was evaluated on the entire trial, studying its characteristics before and after t_{OW}.

Characteristics of Distribution Shift. One of the goals of this paradigm is to better understand how vision systems handle the distribution shifts with different characteristics, such

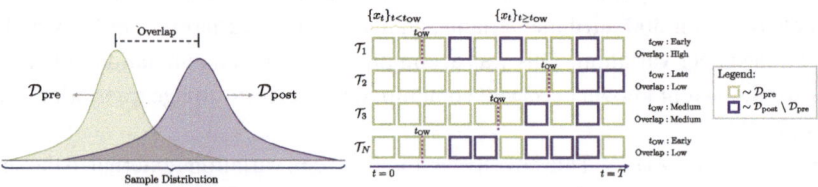

Fig. 4.3 Depicting $\mathcal{D}_{\text{pre}}$ and $\mathcal{D}_{\text{post}}$, potentially overlapping (left). Illustration of different trial scenarios in the benchmark setup (right)

Table 4.1 Estimating difficulty of trials with class and attribute novelty based on sample novelty difficulty and the overlap between $\mathcal{D}_{\text{pre}}$ and $\mathcal{D}_{\text{post}}$

Class/Attribute Novelty		Overlap between $\mathcal{D}_{\text{pre}}$ and $\mathcal{D}_{\text{post}}$		
		High	Medium	Low
Sample Difficulty	Easy	Easy Trials		Hard Trials
	Medium			
	Hard	Medium Trials		

as how does a system respond to the magnitude of distribution shift (or different amounts of overlap between $\mathcal{D}_{\text{pre}}$ and $\mathcal{D}_{\text{post}}$)?

Are some phenomena easier to deal with than others? To assess this, the study also established independent variables that help control for different characteristics during a trial and investigate how a system responds. We discuss these variables below.

- **When Does the Distribution Shift (t_{OW})?**

    Three modes for t_{OW} which controls when the distribution shifts: early ($t_{\text{OW}} \in [0.15T, 0.35T)$), medium ($t_{\text{OW}} \in [0.4T, 0.6T)$), and late ($t_{\text{OW}} \in [0.7T, 0.95T)$) were studied. For a fixed trial length, the early allows systems to see plenty of open-world examples, whereas the late has more samples from $\mathcal{D}_{\text{pre}}$.

- **Overlap Between $\mathcal{D}_{\text{pre}}$ and $\mathcal{D}_{\text{post}}$**

    $\mathcal{D}_{\text{post}}$ is a distributional shift from $\mathcal{D}_{\text{pre}}$, implying that $\mathcal{D}_{\text{post}}$ has closed-world data from $\mathcal{D}_{\text{pre}}$ as well as novelties, i.e., $\mathcal{D}_{\text{post}}$: $\tilde{\mathcal{D}}_{\text{pre}} \cup \{\text{novelties}\}$, where $\tilde{\mathcal{D}}_{\text{pre}} \subseteq \mathcal{D}_{\text{pre}}$. Three modes of overlap were used: high (70–80%), medium (45–65%), and low (0–15%). Low-overlap implies more novel samples, which is harder for the classification task but easier for the detection task, and vice-versa for high-overlap.

- **Difficulty of Novelties**

    This variable allows us to define novelties of three levels of difficulty: easy, medium, and hard, where the difficulty is defined with respect to the classification task. The definition of difficulty varies across the different types of novelties, e.g., a harder class novelty implies it's likely to be confused with a known class, whereas a harder attribute novelty implies confusion with an unknown class, and a harder representation novelty implies difficulty in predicting the correct known class. To estimate the difficulty of class and attribute novelties, features from an independent off-the-shelf baseline were used to compute the average per-class distance to the nearest training classes.

Therefore, for each trial, we select a mode for these independent variables and the type of novelty, along with sample images from $\mathcal{D}_{\text{pre}}$ and $\mathcal{D}_{\text{post}}$; the **difficulty of a trial** depends on the sampled variables. For example, a trial with *easy* class novelty, *late-t_{OW}* and *high-*

Table 4.2 Estimating difficulty of trials with representation novelty based on sample novelty difficulty and the overlap between $\mathcal{D}_{\text{pre}}$ and $\mathcal{D}_{\text{post}}$

Representation Novelty		Overlap between $\mathcal{D}_{pre}$ and $\mathcal{D}_{post}$		
		High	Medium	Low
Sample Difficulty	Easy	Easy Trials		
	Medium	Medium Trials		
	Hard	Hard Trials		

overlap is an *easy trial* for the main classification task, whereas *hard* class novelty, with *early*-t_{OW} and *low*-overlap is a *hard trial*.

Table 4.1 depicts the difficulty of trials for different amounts of overlap and sample difficulties. When the overlap between $\mathcal{D}_{\text{pre}}$ and $\mathcal{D}_{\text{post}}$ is low, irrespective of the sample difficulty, the trial was labeled hard. For medium and high overlaps, the difficulty depended on the sample difficulty. Surprisingly, this was not observed for representation novelty where sample difficulty dictated the difficulty of the trial (Table 4.2).

Overall, we sampled $\sim$1,800 trials, each with 5,000 images distributed across different trial characteristics.

4.4.2 Task Setup

As discussed in Sect. 4.1, we have a main classification task and an auxiliary task of detecting the point of distribution change. We now describe these tasks in detail.

Main Task of Classification. The study's main task is $(K+1)$-way image classification, with K known training classes and an additional unknown class. The goal of an open-world capable system is to maintain the same task performance on $x \sim \mathcal{D}_{\text{post}}$ as on $x \sim \mathcal{D}_{\text{pre}}$, i.e., $\Delta_{\text{Perf}}\left(\mathcal{D}_{\text{pre}}, \mathcal{D}_{\text{post}}\right) = \left|\text{Perf}\left(\{x_t\}_{t \geq t_{\text{OW}}}\right) - \text{Perf}\left(\{x_t\}_{t < t_{\text{OW}}}\right)\right| \sim 0.$

A trivial solution to achieve $\Delta_{\text{Perf}}(\mathcal{D}_{\text{pre}}, \mathcal{D}_{\text{post}}) \sim 0$ is to perform poorly on both distributions. Therefore, an independent, closed-world, **reference baseline**, was developed whose performance was expected to drop in $\mathcal{D}_{\text{post}}$ compared to $\mathcal{D}_{\text{pre}}$, which serves as a reference for several measures.

Auxiliary Task of Detecting Distribution Shift. Assuming that a system is unable to cope with the open-world in a particular trial, it is still of value to know if the system was able to discern that the distribution has shifted [14]. Therefore, we chose **distribution shift detection**, or simply detection, as an **auxiliary task**. Note that good detection performance is not a requirement for good open-world performance on $\mathcal{D}_{\text{post}}$. For example, if an agent's closed-world system is *robust* to the distribution shift in a trial, i.e., there is no performance drop in $\mathcal{D}_{post}$, then *detecting* the shift is not necessary. We refer to this as 'robustness versus detection' trade-off, and emphasize that detection is a diagnostic task.

4.4.3 Results

4.4.3.1 Main Task

In Fig. 4.4, we summarize the classification performance of the agent across different types and levels of novelties. Considering the open-set task, the theory predicts that higher dissimilarity image-level novelty is harder to react to and this is what we observed across *Attribute* and *Representation* novelty where the classification performance degrades, as shown by decreasing $Perf(\mathcal{D}_{post}) - Perf(\mathcal{D}_{pre})$, as the difficulty level increases. There is a slight improvement in performance at *hard class* novelty with respect to *medium class* novelty, which could be attributed to noise in classification of some trials. But overall trends conform to predictions made by the theory.

The theory also predicts that when there is a greater percentage of novel data it will be harder to react to. Finally, it predicts that earlier additions of novelty in the sequence implies more time dealing with the novel distribution, which will be harder to react to. These predictions are confirmed by the observations in Figs. 4.5 and 4.6 respectively.

Fig. 4.4 Change in classification performance $Perf(\mathcal{D}_{post}) - Perf(\mathcal{D}_{pre})$ for Easy (E), Medium (M), and Hard (H) trials

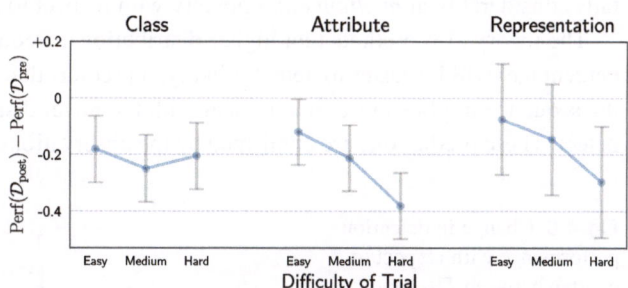

Fig. 4.5 Change in classification performance with respect to overlap between $\mathcal{D}_{pre}$ and $\mathcal{D}_{post}$

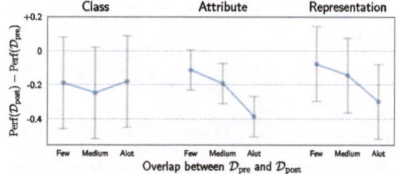

Fig. 4.6 Change in classification performance with respect to time of distribution shift t_{OW}

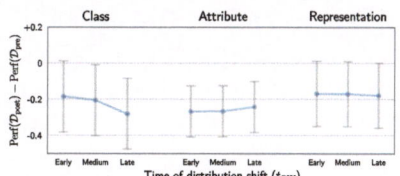

Fig. 4.7 Change in
performance
$\mathrm{Perf}(\mathcal{D}_{\mathrm{post}}) - \mathrm{Perf}(\mathcal{D}_{\mathrm{pre}})$ for
Easy (E), Medium (M), and
Hard (H) trials

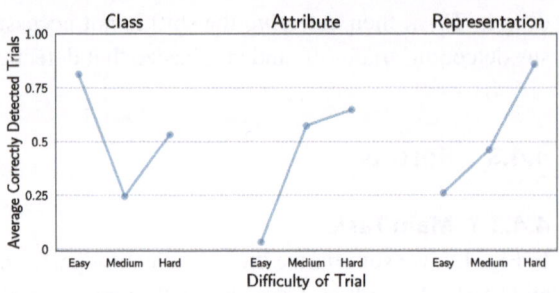

4.4.3.2 Auxiliary Task

Evaluating the auxiliary task of detecting t_{OW} is not as straightforward as classification. If
an agent's detection time step, $\tilde{t}_{\mathrm{OW}}$, is before t_{OW}, i.e., $\tilde{t}_{\mathrm{OW}} < t_{\mathrm{OW}}$, then the detection is
arbitrarily incorrect and such trials have an invalid detection signal. Whereas, if $\tilde{t}_{\mathrm{OW}} \geq t_{\mathrm{OW}}$,
then the detection was delayed by $\delta_{\mathrm{late}} = t_{\mathrm{OW}} - \tilde{t}_{\mathrm{OW}}$ time steps and such trials have a correct
and delayed detection. We refer to the latter as Correctly Detected Trials (CDT).

In Fig. 4.7, we summarize the detection performance of the agent across different types
and levels of novelties. The theory predicts that higher dissimilarity image-level novelty is
easier to detect, and this is what we observed across *Attribute* and *Representation* novelty
where the CDT increases as the difficulty level increases. Surprisingly, the CDT percentage
falls considerably at *medium class* novelty with respect to *easy class* novelty.

The theory also predicts that higher distributional frequency dissimilarity (high novelty
percentage) will be easier to detect. Finally, it predicts that the earlier addition of novelty in
the sequence implies more time to deal with the novel distribution, which will be easier to
detect. These predictions are confirmed by the observations in Figs. 4.8 and 4.9 respectively.

Fig. 4.8 Change in detection
performance with respect to
overlap between $\mathcal{D}_{\mathrm{pre}}$ and
$\mathcal{D}_{\mathrm{post}}$

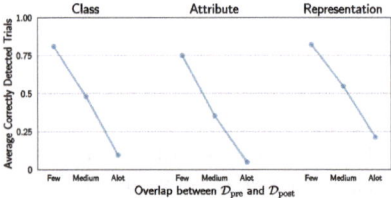

Fig. 4.9 Change in detection
performance with respect to
time of distribution shift t_{OW}

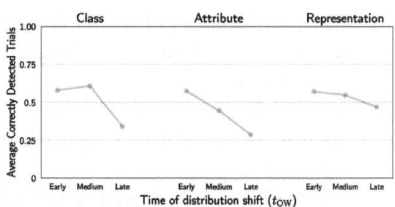

4.5 Conclusions

This chapter applied the unified framework to an image classification task with an auxiliary
task of detecting the change in distribution that occurs when novelty is introduced. The
chapter included a model of difficulty and showed that, as the overall theory predicts, classification system performance, after introduction to novelty, decreased with both difficulty of
the trial and the percentage of novel data. It also showed that for attribute and representation
novelty, the ability to detect novelty increased with difficulty and decreased with overlap in
pre/post distribution and time of the distribution shift.

References

1. Bendale A, Boult TE (2015) Towards open world recognition. In: The IEEE conference on computer vision and pattern recognition (CVPR), pp 1893–1902
2. Jafarzadeh M, Ahmad T, Dhamija AR, Li C, Cruz S, Boult TE (2021) Automatic open-world reliability assessment. In: Proceedings of the IEEE/CVF winter conference on applications of computer vision (WACV), pp 1984–1993
3. He K, Zhang X, Ren S, Sun J (2016) Deep residual learning for image recognition. In: Proceedings of the IEEE conference on computer vision and pattern recognition, pp 770–778
4. Rudd EM, Jain LP, Scheirer WJ, Boult TE (2017) The extreme value machine. IEEE Trans Pattern Anal Mach Intell 40(3):762–768
5. Misra I, Gupta A, Hebert M (2017) From red wine to red tomato: composition with context. In: Proceedings of the IEEE conference on computer vision and pattern recognition, pp 1792–1801
6. Purushwalkam S, Nickel M, Gupta A, Ranzato M (2019) Task-driven modular networks for zero-shot compositional learning. arXiv:1905.05908
7. Dorta G, Vicente S, Campbell NDF, Simpson IJA (2020) The GAN that warped: semantic attribute editing with unpaired data. In: Proceedings of the IEEE/CVF conference on computer vision and pattern recognition, pp 5356–5365
8. Zhu J-Y, Park T, Isola P, Efros AA (2017) Unpaired image-to-image translation using cycle-consistent adversarial networks. In: 2017 IEEE international conference on computer vision (ICCV)
9. Meshry M, Ren Y, Davis LS, Shrivastava A (2021) Step: style-based encoder pre-training for multi-modal image synthesis. In: Proceedings of the IEEE/CVF conference on computer vision and pattern recognition

10. Ehrlich M, Davis L, Lim SN, Shrivastava A (2021) Analysing and mitigating JPEG compression defects in deep learning. In: Proceedings of the IEEE international conference on computer vision workshops
11. Geirhos R, Rubisch P, Michaelis C, Bethge M, Wichmann FA, Brendel W (2018) Imagenet-trained CNNs are biased towards texture; increasing shape bias improves accuracy and robustness. arXiv:1811.12231
12. Mishra S, Shah A, Bansal A, Choi J, Shrivastava A, Sharma A, Jacobs D (2020) Learning visual representations for transfer learning by suppressing texture. arXiv:2011.01901
13. Zabolotny A, Niemann T, Dersch H, German DM, Littlefield KKR, Senore F, Watters J, Rauscher T, d'Angelo P, McKee B, Platt R, Kristian N, Jensen B, de Bruijn P, Modes T, Bronger T, Lensfun library. https://lensfun.github.io/
14. Zhang P, Wang J, Farhadi A, Hebert M, Parikh D (2014) Predicting failures of vision systems. In: Proceedings of (CVPR) computer vision and pattern recognition, pp 3566–3573

Novelty in Handwriting Recognition

5

D. S. Prijatelj, S. Grieggs, F. Yumoto, E. Robertson and W. Scheirer

In the domain of Handwriting Recognition (HWR), many novelties may be encountered and can negatively affect the performance of automatic transcription. This is often the case with historical documents. In the context of this unified framework of novelty, HWR is a task consisting of multiple subtasks. The primary task in HWR is transcription, where novel glyphs, characters, words, and phrases may be encountered during the transcription process. Another key task of HWR is style recognition, including writer identification and overall document image appearance.

5.1 Task Overview

The transcription task involves an agent taking a digital image of a handwritten document as input and processing it to recognize the individual characters to produce a plaintext output. The style recognition task involves the agent identifying known and unknown aspects

D. S. Prijatelj (✉) · S. Grieggs · W. Scheirer
Department of Computer Science and Engineering, Fitzpatrick Hall of Engineering, University of Notre Dame, Notre Dame, IN 46556, USA
e-mail: dprijate@nd.edu

S. Grieggs
e-mail: sgrieggs@nd.edu

W. Scheirer
e-mail: walter.scheirer@nd.edu

F. Yumoto · E. Robertson
PAR Government, 160 Brooks Road, Rome, NY 13441, USA
e-mail: eric_robertson@partech.com

© The Author(s), under exclusive license to Springer Nature Switzerland AG 2024　　　　49
T. Boult and W. Scheirer (eds.), *A Unifying Framework for Formal Theories of Novelty*, Synthesis Lectures on Computer Vision,
https://doi.org/10.1007/978-3-031-33054-4_5

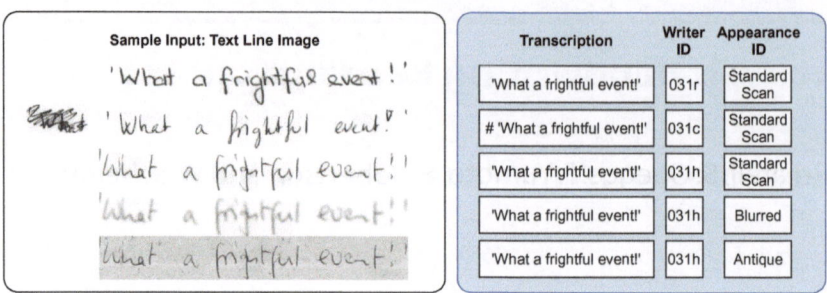

Fig. 5.1 Examples from the IAM offline handwriting dataset [1] depicting the difference in writing style between writers, unknown characters in the scratched out entry represented as "#" in the transcription, and the difference in appearance between the same samples with new modifications to simulate real-world situations. Novelty may occur in any of these labels, such as novel characters, writer, or appearance. HWR agents should be able to handle such novelty

of visual appearance for both the text, e.g., how are individual characters stylized?, and page, i.e., what does the page look like holistically? Two subtasks for style recognition are considered in this paper: (1a) writer identification and (1b) Overall Document Appearance Identification (ODAI). The former involves multi-class classification to distinguish between individual known writers and new writers unseen at training time, while the latter involves multi-class classification to distinguish between known global appearances of handwritten documents and appearances unseen at training time. Any type of novelty can occur in both tasks, thus an important objective of the domain is to detect and manage it. Note that many other tasks can be defined for transcription and style recognition, and the subsequent theory is general enough to cover those as well (Fig. 5.1).

A theory of novelty for the HWR domain can be constructed using the proposed unified framework of novelty in this paper. Recall that a task may consist of multiple subtasks weighted by some priority, making the problem statement above consistent with the notion of a task in the framework, even though it is more complicated than the previous two domains we have discussed.

The theory of novelty for HWR extends the image classification theory of novelty, defined in Sect. 5.4. As is the case for standard image classification, the samples do not necessarily have any meaningful order. However, there is a sequential relation of the characters and words within a sample image of a handwritten document. Throughout this chapter, time step t is in reference to the point in time when a sample image is considered, not a character or word within that image. An HWR task $\mathcal{T}$ can be text transcription, writer identification, or ODAI. A world $\mathcal{W}$ in HWR consists of a d'-dimensional space of pages of handwritten documents. An observation space $\mathcal{O}$ that is accessible to the agent is the d-dimensional space that can encode all possible images of handwritten documents. This space serves as the agent's feature space, extracted from the image. The family of perceptual operators $\mathcal{P}_t$ in HWR are optical sensors, such as cameras, that capture a visible region of the world $\mathcal{W}_t$, where a time step t results in a single image of a handwritten document in the case of still image HWR. The perceptual operator may continue with feature extraction on the captured image to represent the image as a feature vector E_t of arbitrary dimensions to the agent in the observational space. The agent's state space α_T depends entirely on the agent model used. The agent's possible actions are the predictions corresponding to their respective subtask within HWR, such as the transcription text, writer identity or defining style characteristics, or overall document appearance characteristics.

5.2   Dissimilarity and Regret

Given the HWR task within the above world, observation, and agent spaces, the following dissimilarity and regret measures may be better defined:

- The task-dependent world dissimilarity functions $\mathcal{D}_{w,T;E_t}$ and associated novelty threshold δ_w are determined by ground-truth labels associated with the images from the sampling process, where the complete datasets used serve as an oracle. The measurement of dissimilarity uses a distance measure, e.g., Euclidean distance, with a threshold determined by the probability distribution of the data.
- The task-dependent observational space dissimilarity functions $\mathcal{D}_{o,T;E_t}$ and novelty threshold δ_o are determined by the agent's knowledge and design for the task. In a learning-based agent, this is typically done via generalizing from the ground-truth in any available training or validation data. These functions could make use of Euclidean distance or whichever distance measure suits the observational space.
- The world regret function $\mathcal{R}_{w,T}$ is based on the error as measured in the world space given ground-truth labels. For transcription, this could be Levenshtein Edit Distance, Character Error Rate, or Word Error Rate. For the nominal tasks of writer identification and ODAI, the regret is captured by the confusion matrix from which measures of regret derive, such as the Normalized Mutual Information (NMI). NMI tends to be a more desirable summary statistic for the confusion matrix due to ease of interpretability where values near zero indicate poor performance and thus high regret, and values approaching one indicate perfect predictions. NMI also has a strong information theoretic backing and may be theoretically compared to the NMI of random variables with differing sample spaces, such as continuous random variables.
- The observational space regret function $\mathcal{R}_{o,T}$ defines what the agent deems important to the task. This is embodied by the agent's internal model for the task, such as the loss function of a neural network or the likelihood calculation in a probabilistic model.
- The agent's dissimilarity in state is based on the implementation of the agent. The agent's regret is also dependent upon the implementation of the agent. In terms of agents that model the task as an optimization problem, their regret measure would be correlated to the measure used to determine expected errors in current state of the system or the

pairing of that state with an action taken. Thus, the agent regret measure is akin to the loss of a neural network, minimization where the optimum is the minimum, and heuristic functions whose preferred (optimal) direction is also its minimum.

Given the above specification, novelty in HWR is deemed to occur in the world when the world $\mathcal{D}_{w,\mathcal{T};E_t}$ exceeds the novelty threshold δ_w. E_t serves as the history of experience of the agent and plays a key role when a change in the world state is considered novel at a time step. This paper's baseline agent's E_t is simply the training set and indirect information from the validation set, which informs the agent about how it should set its internal threshold δ_o. A world novelty may or may not affect the agent's performance on the task and the novelty's effect may vary in impact to task performance. This is reflected in the world regret $\mathcal{R}_{w,\mathcal{T}}$. The world novelty in any domain, including HWR, must be properly defined in an experiment to assess how an agent performs in its presence.

5.2.1 Caveats of Defining an HWR Oracle

The oracle mainly consists of the datasets used for evaluation. However, the oracle also determines world dissimilarity and regret functions given the data. For datasets that consist of ground-truth labels typically used in supervised learning, the datasets may be enough. However, there may be additional domain knowledge that is not available in the datasets, either implicitly or explicitly defined by the oracle. An example is the case of label frequency within the dataset mismatching the current domain knowledge of the world. In this case, the oracle provides label weightings to adjust the dataset's samples to better represent the current domain knowledge of the universal population of those labels. The information about these weightings may not be provided to the agent as an extended part of the world space.

Another example case is when the dataset has missing labels, either partially or completely, the oracle is expected to provide the information that defines the task given the data. A complete lack of labels for a certain type of information is a rather common occurrence in domains where novelty is present and cannot be labeled in any capacity, and where labeled or unlabeled data may be used in training and evaluation. The datasets do not necessarily define the oracle's information in its entirety. This is a challenge for evaluation design that needs to be accounted for when assessing agents that must manage novelty. A sampling problem, such as HWR, thus defines the oracle and task through the world space, world dissimilarity, world regret, and the information used from datasets and domain knowledge.

5.3   Measurements and Observations

The dissimilarity and regret measures were already specified in the prior section. The observations by space and type have been defined, here we address the samples themselves as they are delivered from the experimental process or world through the perceptual operators to the agent. The observations are the data samples as given over time. Typically, this is done by the experimental process yielding images of handwriting along with their associated labels that define the task, such as the transcription text, writer identity, or style characteristics. This is the typical supervised learning scenario for training an agent. After being trained, the agent will hopefully have learned a generalized model of the task such that new samples given to it through the perceptual operators are able to be properly used to solve the HWR tasks. This includes the agent's detection of when a sample is novel with regards to the task. For example, in transcribing handwriting this would include novel glyphs, characters, words, and phrases. For writer identification, this involves the agent being able to differentiate the given image of handwriting from others and thus be able to differentiate the different style characteristics that make the known writers separable from one another in the observation space. If enough style characteristics in the handwriting differ from all known writers, as understood by the agent, then it will detect novelty. This is accomplished by the agent using a learned detection threshold to specify novelty of a handwriting style profile.

5.4 Novelty Types and Examples

The world dissimilarity and regret functions, respectively, indicate when there is a world novelty, and when it should affect the task. However, as noted in this paper, the occurrence of novelty can be more nuanced and occur in different ways than just world novelty in HWR. These different types of novelty should be addressed and understood when constructing an agent or experiment in HWR with novelty.

World novelty is actual novelty that exists in an environment. For the HWR domain, world novelty, e.g., novel characters, novel writers, novel backgrounds, is the novelty an agent should be most focused on detecting and managing. An observation novelty occurs when the observation from the perceptual operators of the agent is sufficiently dissimilar from every past observation in the agent's stored experience. These novelties can only be detected in the HWR domain if the camera acquiring the image of a document has sensed the novelty that is present in the world, or the agent using the perceptual input detects a novel fault in the camera, such as noise or glitches. Finally, agent novelty occurs when an agent's internal processing cannot map an image to a known state.

Secondary types of novelty exist that are combinations of the above three types. Unanimous novelty is the presence of world novelty, observation novelty, and agent novelty, and represents a valid prediction of novel transcription or style novelty. Imperceptible novelty is a novelty that cannot be sensed by the perceptual operators. In HWR, this can be novelty in the microscopic composition of the material that forms a document page, the historical context in which a document was discovered, the provenance of the document, or any other novelty a camera cannot capture, but a human examiner can determine via other means. Faux novelty is a false positive determination of novelty. False positives can occur if the perceptual operators encounter noise at acquisition time, injecting novelty into the resulting image that does not exist in the environment. Further, in perceptual operators for HWR, imperfect machine learning models for transcription and style recognition have error rates, which can also create a false positive situation.

Regret factors into two additional types of novelty. Managed novelty is a novelty that has a minimal impact on agent performance if it is not detected. In HWR, this could be a change in language expressed by a known character set, which does not impact character transcription performance in any meaningful way. Nuisance novelty is a novelty that is insignificant with respect to world regret but significant with respect to observational space regret. In a document, this could be a stain, tear, or other physical artifact on a page that an agent consistently mistakes for a character (thus negatively impacting its error rate) but has little bearing on the environment of the page from the perspective of the world.

5.4.1   Ontological Specification of HWR with Novelty

An ontological specification serves to describe the knowledge of the world held by the oracle and agent. Functionally, the ontology provides terms and structures to reason about and characterize actual and perceived novelty. We interpret the differences between an agent's task-dependent knowledge of the world and a newly experienced change in the world as a measurable dissimilarity between the world knowledge of the agent and the oracle. The degree of dissimilarity forms a basis for assessing the difficulty an agent has in both detecting novelty and performing its task within that novelty space, which is reflected in the expected world regret $\mathcal{R}_{w,\mathcal{T}}$. For example, in the HWR domain, writing samples from a novel writer with a similar style to a known writer are both difficult to detect as being novel and to identify as being written by an unknown writer.

The ontology's components consist of entities, attributes, actions, relations, interactions (passive), and rules often associated with a specific context or domain. An agent that can detect novelty maintains knowledge elements of the world as described by the ontology. In closed-world supervised learning systems, these knowledge elements are provided through meta-data in the training sets.

The HWR ontology focuses on those components where novelty occurs. We characterize novelty in terms of text elements including writing style and pen selection, as well as in terms of background elements, i.e., those novelties not specific to the text. Writing style corresponds to the writer while the last two correspond to ODAI. The intent of ontological specification is to describe all observable features that may contain novelty including environmental novelties, e.g. water damage to the writing medium, temporal and locale novelties, e.g. date and time representations and document structures, and text-related novelties, such as copyedit marks.

The foundational ontology for transcription and style recognition is shown in Fig. 5.2 (adapted from [2]). We focus on a small group of core attributes representative of each ontological entity that best characterizes the set of novelty in the experiments. We excluded

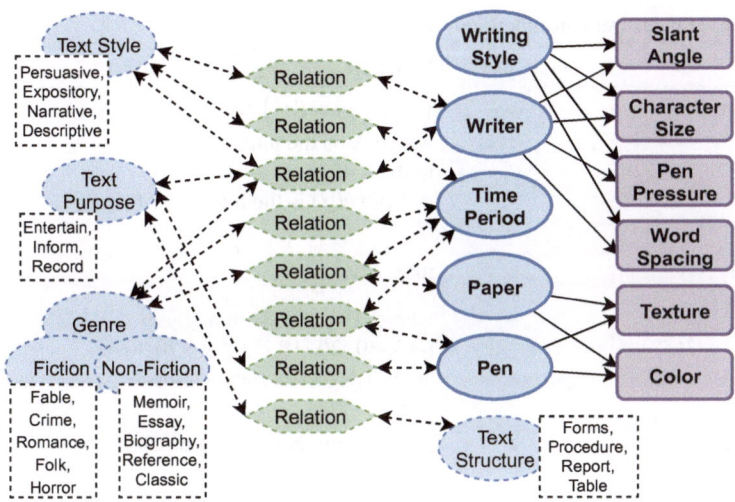

Fig. 5.2 HWR Novelty Ontology adapted from [2]. Entities are represented as blue ellipses. Entity attributes are represented as purple boxes. Examples of entities are dashed white boxes overlapping the entity node. Nodes and edges with solid lines and bold text are the focus of this study

latent attributes from the ontology since they are difficult to qualify in the ontology and beyond the scope of the current study. However, they may play a critical role in novelty detection and characterization.

The HWR ontology defines these entities and attributes:

- Four attributes of writer style: (1) pen pressure, (2) slant angle, (3) character size, and (4) word space.
- The writer with associations to each style attribute.
- The image of a handwriting sample.
- Writing medium (background) categories including types of background noise, textures, and colors.
- Pen categories including textures and colors.

Each component specified by the ontology is associated with a measurement function $\mathcal{F}_{O,c}$ applied to each writing sample in the observation space, where c is a category of novelty. For example, the measurement function for pen pressure is the mean pixel intensity of the written text: $\left(\sum_{i=1}^{N} \text{pixel}[i]\right) / N$, where N is the number pixels in the written text and $\text{pixel}[i]$ is the intensity of a pixel i in the written text. The complete set of measures can be found in Tables 5.1 and 5.2. The collection of normalized measures of each component composes a feature vector for use in distance functions, *e.g.*, cosine similarity, to measure the similarity of a novel writing sample to the body of known non-novel writing samples.

Table 5.1 Style measurement functions

Style	Function
Pen Pressure	$\left(\sum_{i=1}^{\mathcal{N}} pixel[i]\right)/\mathcal{N}$ where $\mathcal{N}$ is the number pixels in the written text, and $pixel[i]$ is the intensity of a pixel i
Slant angle	$\max_{A^i} \; S(A^i)$ where A^i is the set of angles [-45,-30,-20,-15,-5,0,5,15,20,30,45], and $S(A^{(i)})$ is a shear estimate [4]
Word spacing	Average number of horizontal pixels between words where a space is a vertical slice with fewer than 30% quantile of vertical pixels for a line image
Character size	Average number of pixels over all vertical slices of the image excluding those slices labeled as a space

Table 5.2 Non-style measurement functions

Novelty type	Function
Background	Entropy of the grey level background (without text)
Pen	Entropy of the grey level pixel intensities in the written text

   We can also represent the writing style attributes graphically by first creating discrete attributes through binning the component measures and assigning each style and bin to a node in the world knowledge graph, with an example in Fig. 5.3. Similarity of writing styles is represented within the knowledge graph by the shared style relations. The graph supports the application of graph metrics such as isomorphism between two writing style sub-graphs, where a higher number of shared discrete attributes between writing styles indicates a higher similarity. We hypothesize that the degree of dissimilarity inversely impacts the ability of an agent to detect and characterize novelty. However, the choice of entities, fidelity for measurement for each ontological entity, and weight of significance for each entity impact the utility of dissimilarity measures as discussed below in Sect. 5.5.2.

Fig. 5.3 Illustrative
Knowledge graph of Writing
Style for four style attributes
associated writing samples
a05-022-07 and r06-143-01
and five selected writers. The
red edge on the bottom
represents the writing style of a
sample not associated with the
sample's author

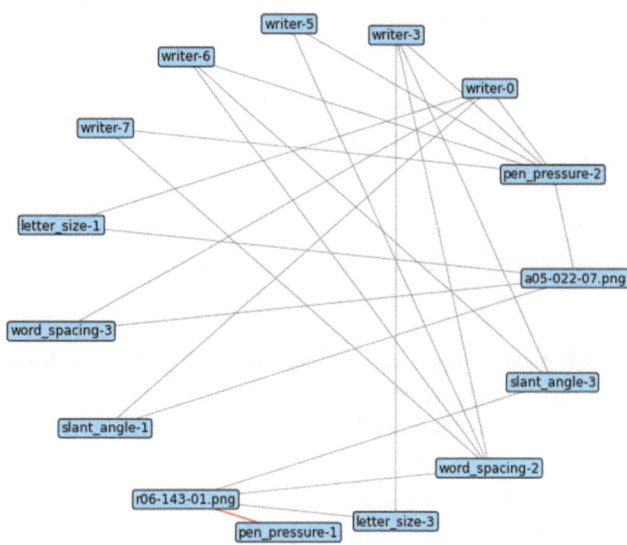

In the development of the ontology-based knowledge graph, we first start with the characterization of writing style. Writing style is made up of the style attributes slant angle, word space, character size and pen pressure. Each style attribute is described by a continuous function, defined in Table 5.1, which is applied to images of words present in each writing sample [3]. The results from the functions for all samples are binned to form discrete style descriptors, which are used to construct style attribute nodes in the knowledge graph. The number of bins is chosen to provide adequate separation of each writing style. In our initial assessment, we used four bins for slant angle and three bins for the rest of the style attributes. The style attributes are collected for all writing sample images. The most frequent style attribute value is assigned to each writer. The result is a set of associations between each writing sample, the style, and the writer, as shown in the knowledge graph in Fig. 5.3. We apply the same approach for background and pen novelties. The non-style measurement functions for these novelties are described in Table 5.2.

A correct knowledge graph consists of each writing sample associated to a single writer via a two-step path through the four style attribute nodes. The writing style measurement functions provide a gross measure of writing styles. Combined with the inaccuracies introduced with binning, not all writing samples from the same writer are associated with the same set of bins across all style attributes. Since the same writer's style is an aggregate value over a set of writing samples, some binned measures for a writing sample form an association to a style attribute not associated with the writer of the sample. This is highlighted in the sample graph in Fig. 5.3 via a red edge. This suggests an optimization strategy for style binning and association to maximize the number of writing samples associated with a writer through the four style attribute nodes.

Fig. 5.4 Example Clusters for
two cluster groups, Style and
Background, under three
novelty types: (1) No Novelty;
(2) Novel Style; and (3) Novel
Background

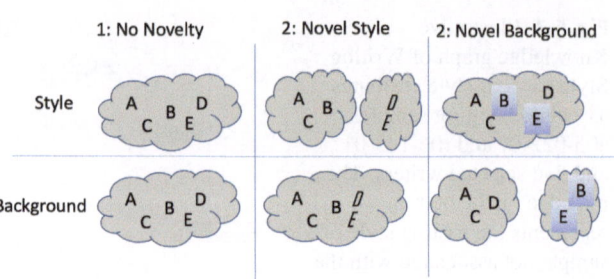

Characterization was achieved through groups of clusters over writer samples created by
the agent. Each group explains a single characterization of novelty as it occurs in each text
image. Groups included in our initial study are:

- Up to three clusters for pen pressure, character size, and word spacing,
- Up to four clusters for slant angle,
- Up to three clusters for category of novelty: writer novelty, background, and pen novelties.

A single 'writer novelty' cluster occurs in the novelty category cluster group when novelty
does not occur — all non-novel examples cluster together. Figure 5.4 illustrates this approach
with two cluster groups.

For performance evaluation of characterization, we use Normalized Mutual Information
(NMI) to measure the quality of the clusters. We first separate the agent characterizations of
writing samples with no novelty and the three categories of novelty: writing style, pen, and
background. We interpret characterizations in the non-novel subgroup as a base measurement
of the agent's dependence on the cluster-represented attributes to describe novelty.

The characterization promotes better understanding of an agent's performance in the
HWR domain with novelty. We first establish a baseline cluster quality using non-novel
writing samples. In the baseline, cluster groups organize samples by similar styles and
backgrounds. As different types of novelty are introduced, new cluster centers are formed
to isolate those samples perceived as having the group's representative novelty.

We partition and evaluate the characterization clusters by novelty category, shown in each
row of Table 5.3, to highlight the interactions between different categories of novelty and the
novel style attributes. Applicable measures to this structure include NMI and cluster purity.
This structure for analysis aids in understanding agent response to mixed novelties, such as
style and background changes. The "No Novelty" row serves as a baseline characterization
of writing samples without novelty. The Style row measures characterization clusters of
samples with novel writing styles. In terms of a mapping to empirical observations, low
performance for cell PP_p (Pen, Pen Pressure) in comparison to baseline No Novelty was
observed, indicating that this paper's open-world agent is unable to discern pen changes
from pen pressure novelty. We do not expect to see much variation from the non-novel

Table 5.3 Characterization cluster groups are Pen Pressure (PP), Character Size (CS), Word Spacing (WS), Slant Angle (SA), and Novelty Category (NC)

Novelty	PP	CS	WS	SA	NC
Style	PP_s	CS_s	WS_s	SA_s	NC_s
Background	PP_b	CS_b	WS_b	SA_b	NC_b
Pen	PP_p	CS_p	WS_p	SA_p	NC_p
No Novelty	PP_n	CS_n	WS_n	SA_n	NC_n

examples in cells CS_p, WS_p and SA_p, since pen pressure novelties do not significantly affect character size, slant angle, and word spacing. We also expect matched performance to the baseline No Novelty conditions for separable novelty categories, such as all Style cells in the Background row (PP_b, CS_b, WS_b, SA_b). For example, an agent is expected to separate novel from non-novel backgrounds, but may fail to adequately separate groups of samples using two different novel backgrounds.

The Novelty Category (NC) cluster group serves to characterize the core types of novelty. NC cells NC_c, NC_b, and NC_p are meant to be measurements of an agent's ability to distinguish different samples within the same category of novelty. For example, NC_b measures an agent's ability to distinguish examples with blue backgrounds from those with red backgrounds.

For an initial assessment, we characterized the last 32 test images selected from each test prior to evaluating characterization of the novelty. We provide the sample set of measurements using Cluster Purity in Table 5.4,

$$\text{Purity} = \frac{1}{N} \sum_{i=1}^{k} max_j |c_i \cap t_j|, \tag{5.1}$$

where N = number of writing samples, k = number of clusters, c_i is a cluster in C, and t_j is a ground-truth novelty label.

Table 5.4 Cluster purity characterization results based on novelty type. Characterization cluster groups here are Pen Pressure (PP), Letter Size (LS), Word Spacing (WS), Slant Angle (SA), and Novelty Category (NC)

Novelty	PP	LS	WS	SA	NC
Style	0.88	0.85	0.55	0.53	1.00
Background	0.83	0.89	0.45	0.61	0.89
Pen	0.75	0.71	0.57	0.77	1.00
Non-Novel	0.84	0.75	0.80	0.80	1.00

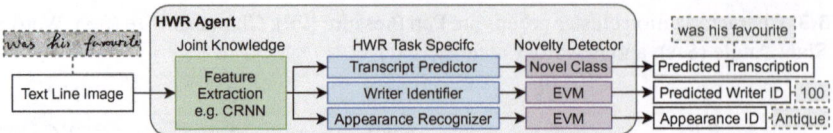

Fig. 5.5 The baseline open-world HWR agent from [5] that is able to detect novelty consists of modules that leverage joint and task-specific knowledge of the HWR domain. Thus, it performs the tasks as desired while managing novelty. The "Appearance Recognizer" is the module for identifying the global appearance of the document, e.g., white background with black foreground, noisy, or antique. The "Novel Class" novelty detector for transcription indicates that the model is trained with a class that represents novelty to isolate those instances from known classes

In this sample, we see evidence of confounding variables when characterizing pen pressure with pen changes. Characterization of slant angle, when compared to non-novel cases, is weakly affected by pen changes. Word spacing is significantly affected in all three novelty cases. Style changes are correctly separated from background and pen novelties, as indicated in the Novelty Category cluster group (Fig. 5.5).

5.5   Experiments

Multiple experiments were used to apply and analyze the novelty framework to the domain of HWR. A large-scale test using 55,000 samples evaluated a baseline agent designed to handle novelty in the HWR domain.

Two baseline HWR agents were used, taken from [5], an agent that was designed to handle only a closed-world and another designed to handle an open-world. The baseline open-world HWR agent consisted of a CRNN for transcription and a pair of EVMs given the mean Histograms of Oriented Gradients (HOG) of the images as input for the style tasks of writer identification and ODAI. The CRNN's sequential neural network structure is depicted in Table 5.6, read in top-down order, and a high level depiction is in Fig. 5.6. A closed-world agent was evaluated as comparison points to the open-world agent. The closed-world agent's sequential neural network structure is seen in Table 5.5, also read in top-down order. The closed-world agent performed the writer identification style subtask and the text transcription task. The agent did not pass information between each of its tasks and had no specific abilities to manage novelty.

For writer identification, the closed-world baseline agent predicted, for each sample, one of the 50 known writers in the training set by applying the SoftMax function to the output of the dense layer of a CNN. This closed-world model served to demonstrate limited utility only in a closed-world, over-fit to known writers, with considerable degradation in performance when exposed to novel conditions. The closed-world writer identification model was a neural network consisting of three groups of 2D convolution layers with RELU activation and max pooling, followed by two groups of dense connected layers with RELU activation and

Table 5.5 Baseline closed-world writer-predictor agent model	zero_padding2d	(115, 115, 1)
	Conv-2D	(58, 58, 32)
	MaxPooling2D	(29, 29, 32)
	Conv-2D	(29, 29, 64)
	MaxPooling2D	(14, 14, 64)
	Conv-2D	(14, 14, 128)
	MaxPooling2D	(7, 7, 128)
	Flatten	6,272
	DropOut	6,272
	Dense	512
	DropOut	512
	Dense	256
	DropOut	256
	Dense	50

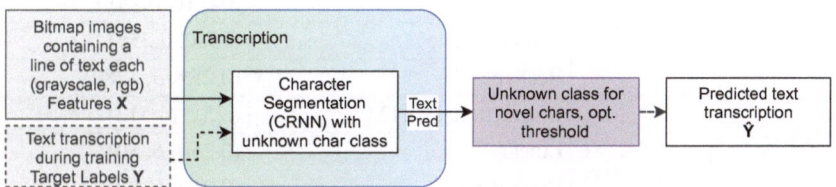

Fig. 5.6 The CRNN in isolation depicting its joint use as both the feature extractor and transcript predictor. The feature extraction occurs in joint together with its supervised learning of the transcription task. The penultimate layer of the CRNN is used as one of the examined feature spaces for training the style task EVMs, after using PCA on the zero padded sequential output. Due to memory constraints, 1,000 components were used after reduction with an incremental implementation of PCA to on 25% of the training data

50% drop out, ending with a dense SoftMax activated layer over the 50 known writers [6]. Figure 5.5 depicts the sequential network architecture. For transcription, the closed-world agent used a deep recurrent network with log-SoftMax outputs [7].

The mean HOG configuration of the baseline open-world HWR agent was evaluated with 55,000 tests. We generated 55,000 tests based on experimental conditions. For each generated test, we created nine additional tests, re-ordering the test samples to average-out sample variations while retaining the same conditions of the test. Tests were constructed and grouped by types of novelty. The tests were constructed to evaluate both single writing sample novelty detection and world change detection indicated by data distribution change from a non-novelty phase to a novelty phase of the test. In addition to establishing a foundation for novelty detection and characterization in HWR, in this evaluation, we established some initial

Table 5.6 Baseline open-world agent. x is input size, b is batch size. First column indicates repeated sequential blocks of layers

	Conv2d	$(16, 64, x)$
	BatchNorm2d	$(16, 64, x)$
	LeakyReLU	$(16, 64, x)$
	MaxPool2d	$(16, 32, 0.5x)$
	Conv2d	$(32, 32, 0.5x)$
	BatchNorm2d	$(32, 32, 0.5x)$
	LeakyReLU	$(32, 32, 0.5x)$
	MaxPool2d	$(32, 16, 0.25x)$
x2	Conv2d	$(48, 16, 0.25x)$
	BatchNorm2d	$(48, 16, 0.25x)$
	LeakyReLU	$(48, 16, 0.25x)$
	Dropout2d	$(48, 16, 0.25x)$
	Conv2d	$(64, 16, 0.25x)$
	BatchNorm2d	$((64, 16, 0.25x)$
	LeakyReLU	$(64, 16, 0.25x)$
	Flatten Interior	$(1280, 0.25x)$
	Reshape	$(0.25x, b, 1280)$
x5	Bidirectional LSTM	$(.25x, b, 512)$
	Linear	$(0.25x, b, 80)$
	LogSoftmax	$(0.25x, b, 80)$

metrics for novelty difficulty, identifying factors impacting the performance of detecting novelty, and transcribing handwritten text.

5.5.1 Protocol: Modified IAM Off-Line Handwriting Data

The roughly 55,000 novel writing samples used in evaluation were constructed from modified samples of the IAM Offline Handwriting Dataset [1]. The training data will be publicly released after this paper's publication. A representative portion of the tests will be released as well.

Training and evaluation data, in the form of individual lines, were selected from IAM. Prior to training, lines were denoised, removing shadow boxes around the letters of each word. Features were then extracted from the clean lines of written text to capture writing characteristics including pen pressure, letter slant, word spacing, and character size [8]. A distance matrix was formed by the sum of absolute differences between each writer's mean style across all example words from each writer. This distance matrix served as a writer similarity measurement.

The training set was made up of lines of text from 50 selected writers representing a subset of the writer's style descriptor values, leaving one or two bins for each feature excluded for use in the novelty evaluation set. Lines of text did not contain any additional effects, using a white background. Sample lines from six additional writers were chosen to compose an unknown writer training set. The set was supplemented with samples from the RIMES dataset and samples from the same 50 writers with background effects including salt and pepper noise, antique paper, and faded impressions of shaded boxes around the words in each line of text.

The evaluation set was made up of the remaining writers and writing sample manipulations to alter characteristics of both the writing style and the background. Letter style manipulations included thinning or widening, brightness, resizing, and slant adjustments to each line of text. The background was composed from Creative Commons licensed images of textured paper. Pen manipulations were similarly constructed by merging in textures and colors, weighted by the pixel strength, i.e., pen pressure.

The difficulty associated with each test was determined by the novelty type. The difficulty for novel writers and novel letter manipulations was determined by the ontological separation of four writing style features: pen pressure, letter slant, character size, and word spacing. Grouping novel writers with non-novel writers with similar styles is intended to make detection more difficult. The difficulty of novel pen and backgrounds was measured by the inverse intensity of the background (since the letters are black).

Most novel examples were constructed with a single type of novelty. Background and pen novelties were applied to sample text lines from the 50 known writers. The number of text lines per test varied based on availability of data targeting the specific novelty: 512, 768, or 1,024. In total, each test selected from 1,696 non-novel examples of the 50 known writers and approximately 50,000 novelties. Tests were composed of writing samples selected and organized by six independent discrete variables defining the experimental conditions of each test to explore the performance regime in novelty detection, resulting in 3,888 unique combinations. Using several subtypes, e.g., different backgrounds, of novelties by type and difficulty, we constructed approximately 5,500 tests, each reordered nine times to average out sample variations.

Difficulty and novelty type (Table 5.7) affect writer prediction and transcription accuracy. Variables (Table 5.8) associated with distribution and placement of novelty in stream of data, such as introduction point, density of novel to non-novel samples, and distribution type were varied to measure impact on novelty detection.

Table 5.7 Number of writing samples for each type of novelty

Novelty type	Count
Background	17,662
Letter	11,868
Pen	11,289
Writer	8,427
No Novelty	1,696

Table 5.8 Independent variables forming the experimental conditions of each novelty test

Independent Variables	Values
Mean novelty Introduction Pt.	0.4, 0.5, 0.6, 0.7, 0.8, 0.9
Density of novelty	6 different densities
Novelty type	Writer, Letter, Background
Difficulty	Easy, Medium, Hard
Distribution type	High (positive skew), Low (negative skew), Mid (normal), Flat (uniform)
Test length	512, 768 and 1,024

5.5.2 Large-Scale 55K Evaluation Results

This section provides a fine-grained analysis over the 55,000 tests presented to the closed-world agents and the novelty detecting open-world agent described above. The agent configuration used for the open-world agent experiments utilized HOG features for all style tasks. For this analysis, we present general novelty detection, text transcription, and writer identification performance across all tests based on types of novelty.

5.5.3 Closed-World Agents: Transcription and Writer Identification

As expected, novelty negatively impacted the writer identification and sample transcription accuracy (results are shown in Table 5.9). Mean character transcription accuracy was reported as $1 - L(G_s, A_s)/\max(|G_s|, |A_s|)$ where L is Levenshtein Edit Distance, G_s is ground-truth text for writing sample s, and A_s is the agent's predicted transcription for writing sample s, averaged over the ten variations of each test. Writer identification accuracy was reported as mean accuracy of the top-1 and top-3 predictions out of $K+1$ writers, where K = 50 for all tests, and the additional class was for novel writers.

Table 5.9 Baseline closed-world agent mean character transcription accuracy, top-3 writer identification accuracy, and top-1 writer identification accuracy in response to non-novel and novel writing samples

Is Novel?	Mean Char. Acc.	Writer ID Top-3 Acc.	Writer ID Top-1 Acc.
False	0.85	0.99	0.99
True	0.47	0.40	0.24

Table 5.10 Baseline open-world agent mean character transcription accuracy, top-3 writer identification accuracy, and top-1 writer identification accuracy in response to non-novel and novel writing samples

Is Novel?	Mean Char. Acc.	Writer ID Top-3 Acc.	Writer ID Top-1 Acc.
False	0.82	0.942	0.719
True	0.62	0.479	0.220

5.5.4 Open-World Agent: Novel Verses Non-novel Predictions

Again, as expected, novelty negatively impacted both text transcription and writer identification accuracy. However, the open-world agent was significantly better at the text transcription task (results are shown in Table 5.10). Transcription performance was reported as mean character accuracy computed using the ground-truth and the agent provided transcriptions for all tests. Writer identification accuracy was reported as mean accuracy of the top-1 and top-3 predictions out of $K + 1$ writers, where $K = 50$ for all tests.

5.5.5 Open-World Agent: Novel Style Manipulations

Style manipulations included manipulations to the characters. These manipulations had a measurable impact on writer identification performance (results are shown in Table 5.11). Dilating the letters did not affect performance, down-weighting pen width as a major factor of a writer's style. More extreme character manipulations, such as large slants and slants coupled with dilation, were more easily detected as being novel, as expected. Inverting pixel values for written text did not adversely affect writer identification performance. The novelty detector did not equate letter inversion as novelty. Each novelty type was represented by 1,696 sample images.

Four different summary statistics were computed for the novel style manipulations. Novelty detection accuracy was mean accuracy of all of the detection decisions. Mean character transcription accuracy is defined as,

$$1 - L(G_s, A_s)/\max(|G_s|, |A_s|), \tag{5.2}$$

Table 5.11 Novelty detection accuracy, mean character transcription accuracy, top-1 writer identification mean normalized mutual information, and top-3 writer identification accuracy given pen novelties grouped by novel style changes

Novelty type	Novelty Detect. Acc.	Mean Char. Acc.	NMI	Writer ID Acc.
Dilate	0.99	0.70	0.01	0.57
Erode	0.79	0.77	0.35	0.68
Increase size	0.99	0.33	0.01	0.02
Big right slant	0.79	0.62	0.21	0.09
Slant w/Dilate	0.99	0.46	0.04	0.00
Big left slant	1.00	0.55	0.01	0.02
Small slant	0.86	0.52	0.23	0.04
Inverted	0.33	0.71	0.79	0.94

where L is Levenshtein Edit Distance, G_s is ground-truth text for writing sample s, A_s is the agent's predicted transcription for writing sample s, averaged over the ten variations of each test. NMI represents normalized mutual information between the actual writer of the sample and the top-1 predicted writer. Writer identification accuracy is mean accuracy of the top-3 predictions out of $K+1$ writers, where $K = 50$ across all tests, and the additional class is for novel writers. These summary statistics are also used for the novel pens and novel backgrounds assessments, which are described below.

5.5.6   Open-World Agent: Novel Pens

Novel pens included manipulations to written text, replacing the pixels with textures and colors, weighted by the intensity of the pen as described by pen pressure (results are shown in Table 5.12). Pen manipulations had minimal impact on writer identification performance. Each novelty type was represented by 1,696 sample images.

5.5.7   Open-World Agent: Novel Backgrounds

Background manipulation had a more diverse impact on writer prediction performance than style manipulations (results are shown in Table 5.13). NMI represents Normalized Mutual Information between the actual sample writer and top-1 predicted writer. Novel types of shadow boxes (from the uncleaned lines extracted from IAM) had the highest writer identification accuracy, perhaps due to similar associations made with these types of artificial irregularities in the training set. As with pen manipulations, more extreme manipulations

Table 5.12 Novelty detection accuracy, mean character transcription accuracy, top-1 writer identification mean normalized mutual information, and top-3 writer identification accuracy grouped by novel pens

Novelty type	Novelty Detect. Acc.	Mean Char. Acc.	NMI	Writer ID Acc.
Blue color	0.98	0.78	0.09	0.53
Brown texture	0.74	0.75	0.43	0.79
Gold texture	0.92	0.69	0.22	0.73
Rainbow	0.98	0.71	0.10	0.57
Red color	0.98	0.70	0.10	0.53

Table 5.13 Novelty detection, mean character transcription accuracy, top-1 writer identification mean normalized mutual information, and top-3 novel writer identification accuracy grouped by writing style

Novelty type	Novelty Detect. Acc.	Mean Char. Acc.	NMI	Writer ID Acc.
Antique	0.40	0.80	0.47	0.39
Blue fabric	0.98	0.40	0.10	0.42
Blue color	0.98	0.69	0.01	0.54
Blue wall	0.99	0.33	0.01	0.10
Brown fabric	1.00	0.60	0.01	0.11
Crinked paper	1.00	0.71	0.02	0.16
Gaussian noise	1.00	0.14	0.00	0.16
Gold wall	0.68	0.51	0.34	0.37
Rainbow paper	0.98	0.25	0.12	0.54
Shadow boxes	0.59	0.88	0.62	0.91

resulted in higher detection accuracy. Increased texture interfered with the agent's ability to identify the writer.

5.5.8   Open-World Agent: Writer Similarity in Novel Writer Discovery

Each test was composed of sample writing from known and unknown writers. Here we found the minimum distance of an unknown writer across all known writers. We hypothesize that the greater the distance of writer style attributes of unknown writers with known writers, as captured in the ontological specification, the easier it is to detect a novel writer.

Table 5.14 Novelty detection and novel writer identification correlation grouped by writing style

Style	Novelty Det. Corr.	Writer ID Corr.
Slant angle	0.06	0.01
Skew angle	0.02	0.04
Word spacing	−0.02	-0.05
Pen pressure	0.14	0.09
Character size	0.03	0.04
Summed	0.12	0.18

Surprisingly, the results did not show a strong correlation as expected. Table 5.14 shows the Pearson's correlation of each style attribute with detection and top-1 novel writer identification accuracy. We believe this is due to two key factors: not enough variability in writing styles in the unknown population and the chosen set of attributes insufficiently capturing all of the essential characteristics of writing style. Pen pressure had the highest correlation of the four ontological specified factors. Collectively, a weak positive correlation did support the hypothesize. The proposed benchmark can be augmented with additional attributes as the challenge problem evolves.

5.5.9 Open-World Agent: Factors in Novelty Detection

A critical factor in the 55,000 tests was the density and location of novelty introduction—the switch between pre-novelty and post-novelty phases of the test given a stream of writing samples. This approach treats novelty as perceived world changing events rather than outliers, where confidence of novelty predictions increases as more novel examples are encountered in the data stream, increasing the body of evidence. With this approach, the level of false positives, those misidentified non-novel examples that fall in the pre-novelty phase of the test, can be substantially reduced. Figure 5.7 shows the false positive count by the proportion of novelty. The variability and amount of false positives decreased as the proportion of novel samples to non-novel samples increased.

We conducted ANOVA to identify factors affecting the false positive rate (see Table 5.15). Along with the proportion of novelty, distribution type had a significant impact on the false positive rate. A positively skewed distribution, where novel samples densely occurred at the start of the novelty phase of the test, was associated with a lower false positive rate when compared to other distribution types, such as a negatively skewed distribution. Novelty difficulty had a weak association to the false positive rate.

Fig. 5.7 False positive count of HWR agent by the proportion of novelty present indicating stability in false positive rate as novelty increases

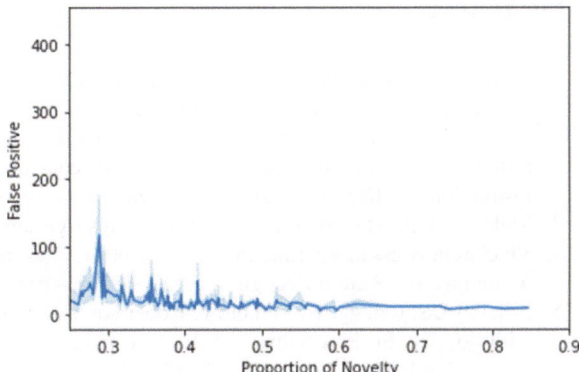

Table 5.15 ANOVA analysis of statistical influence given several test generating independent variables identified in Table 5.8 on false positive rate

Factor	Sum of squares	df	F	p
Distrib. type	1.058e+06	3	129.300	0.000
Level of difficulty	7.456e+03	2	1.365	0.255
Location of novelty	2.646e+06	1	969.301	0.000
Proportion of novelty	5.848e+05	1	214.240	0.000
Residual	1.205e+00	44170		

5.6   Conclusions

This chapter introduced the problem domain of novelty in handwriting recognition (HWR). This domain is attractive for the study of novelty, as it consists of a key challenge problem within AI: reading in a more human-like way. The HWR domain with novelty was formalized, an evaluation protocol with benchmark data was introduced, and comprehensive results from a baseline agent were presented to provide the research community with a starting point to build upon. Beyond incremental improvements in transcription performance and style recognition in the presence of novelty, we suggest that adaptation via incremental learning is the next step. Agents that can properly react to and manage novelty, as opposed to merely detecting novelty, will perform better on the task over time. With additions to the evaluation protocol supporting this, we expect a new class of agents to appear for a number of document processing applications.

References

1. Marti U-V, Bunke H (2002) The IAM-database: an English sentence database for offline hand-writing recognition. Int J Doc Anal Recognit 5(1):39–46
2. Ontario Ministry of Education (2006) A guide to effective literacy instruction, grades 4 to 6: a multivolume resource from the ministry of education. Volume one, Foundations of literacy instruction for the junior learner, p 37. Ontario Ministry of Education
3. Malemnganba M (2018) ml-graphology. https://github.com/Malemm/ml-graphology
4. Vinciarelli Alessandro, Luettin Juergen (2001) A new normalization technique for cursive hand-written words. Pattern Recognit Lett 22(9):1043–1050
5. Prijatelj DS, Grieggs S, Yumoto F, Robertson E, Scheirer WJ (2021) Handwriting recognition with novelty. In: Proceedings of the 16th international conference on document analysis and recognition ICDAR 2021, ICDAR 2021. Springer lecture notes in computer science
6. Reddy T (2018) Offline handwriting recognition cnn. https://github.com/TejasReddy9/handwriting_cnn
7. Shi B, Bai X, Yao C (2015) An end-to-end trainable neural network for image-based sequence recognition and its application to scene text recognition
8. Joshi PM, Agarwal A, Dhavale A, Suryavanshi R, Kodolikar S (2015) Handwriting analysis for detection of personality traits using machine learning approach. Int J Comput Appl 130:40–45, 11

Contextual and Semantic Novelty in Text

6

N. Ma, B. Liu and E. Robertson

Novelty, anomaly, or out-of-distribution (OOD) detection has been an active research area for decades, since the 1960s [1, 2], due to its widespread applications in various domains [3, 4], such as financial surveillance, health and medical risk, AI safety, network intrusion detection, etc. Recently, this research has also received increased attention in the Natural Language Processing (NLP) domain. For example, many researchers have studied the problem of text classification [5–9], but, as opposed to classical text classification that works under the closed-world assumption, novelty or OOD detection works under the open-world setting, which *allows the class of a test instance to be unknown*. However, the existing text classification methods that perform novelty or OOD detection are mainly ***coarse-grained*** and ***topic-based***. Given a text document, their goal is to detect whether the text belongs to a known topic or an unknown topic. Some examples include detecting product reviews of unknown product categories [7], news articles of unknown topics [6], and natural language commands of unknown intents/domains [9]. This chapter is based on two papers published in 2021 and 2022 [10, 11]. They introduce a new novelty detection problem in the NLP field - *fine-grained semantic novelty detection in texts*, which we also call *contextual novelty detection*. In this chapter, we want to present a learning-based agent for semantic novelty detection in text. The agent does not actively interact with the world. The perceptual operators are Deep Neural Networks (DNN) built on text features created from large language models and external knowledge bases.

N. Ma · B. Liu (✉)
Department of Computer Science, University of Illinois Chicago, Chicago, IL, USA
e-mail: liub@uic.edu

E. Robertson
PAR Government, Rome, USA

© The Author(s), under exclusive license to Springer Nature Switzerland AG 2024
T. Boult and W. Scheirer (eds.), *A Unifying Framework for Formal Theories of Novelty*, Synthesis Lectures on Computer Vision,
https://doi.org/10.1007/978-3-031-33054-4_6

6.1   Task Overview

In this chapter, we use the terms *semantic* and *contextual* interchangeably. Given a text description x in Natural Language (NL), we detect whether x represents a semantically novel phenomenon or not. In our daily lives, we observe different real-world phenomena, e.g., events, activities, and situations, and often describe these phenomena in NL to others or write about them. It is quite natural to observe scenes, activities, or factual texts that we have not seen before and are novel to us. In this chapter, we study two types of text x. In the first type, x is a NL scene description (such as x_1 and x_2 shown in Fig. 6.1), which can be the caption of an image/video clip or a piece of scene description in a novel/poem. It is a common scene that "*A person **walks a dog** in the park*," but if someone says "*A man is **walking a chicken** in the park*," it is quite unexpected and novel. In the second type, x is a factual text involving two *named* entities of interest, such as x_3 and x_4 shown in Fig. 6.1. Factual texts appear in all kinds of media, such as news articles, blog posts, and even reviews, which often contain many semantically novel sentences involving popular real-world (named) entities.[1] Given a factual text x containing two *named entities*,[2] the goal is to classify whether x represents a semantically novel fact or a normal one *with respect to* the entity pair. For example, consider the text x_3 and an entity pair underlined in x_3 in Fig. 6.1. x_3 represents a *normal* fact as it is natural for an actor (*Johnny Galecki*) to act in the sitcom or TV show (*The Big Bang Theory*). However, x_4 in Fig. 6.1 depicts a novel fact with respect to the underlined entity pair because a CEO of a technology company (*Elon Musk*) acting in the sitcom (*The Big Bang Theory*) is quite surprising and novel.

An important application domain of our proposed task is the development of engaging software. In the Internet era, with all kinds of mobile applications, designing an application that is not only usable but also attractive to users is vital. If an application fails to attract users' attention and engagement, users may lose interest in the application and move to other applications [12, 13]. Developers and researchers have focused on designing products with non-utilitarian features to promote more user engagement [14, 15]. Our proposed task is important for developing engaging software because novel phenomena triggers users' interest and naturally appeal to most users who innately want to satisfy their curiosity. For example, a newsfeed mobile application can increase user engagement by recommending novel news/facts of the named entities of interest, promoting news articles with novel facts appealing to users' curiosity, etc. A chatbot is more intelligent and engaging if it can detect novel text from a user in a conversation and also reply with novel scene description, factual text, or images and videos based on text descriptions and context information.

Problem Definition. In this chapter, we address the problem of *fine-grained semantic novelty detection in texts* [10, 11]. Given a set of natural language descriptions $\mathcal{X} = \{x_1, x_2, \dots x_n\}$ of common/normal text, build a model $\mathcal{M}$ using $\mathcal{X}$ to score the semantic novelty of a test NL description of a phenomenon x' with respect to $\mathcal{X}$, i.e., classifying x'

---

[1] We use the term *entity* and *named entity* interchangeably.

[2] Named Entity definition: https://en.wikipedia.org/wiki/Named_entity.

Fig. 6.1 Examples of semantic novelty detection in texts

x_1 :	"*A person **walks a dog** in the park.*"	normal
x_2 :	"*A man is **walking a chicken** in the park.*"	novel
x_3 :	"*The Big Bang Theory is an American television sitcom, filmed in front of a live audience, stars Johnny Galecki et al.*"	normal
x_4 :	"*Elon Musk's performance as a dishwasher in a restaurant in season 9, episode 9 of the The Big Bang Theory is quite interesting to his fans.*"	novel

into one of the two classes {*NORMAL, NOVEL*}. "*NORMAL*" means that x' is a description of a common phenomenon and "*NOVEL*" means x' is a description of a semantically novel phenomenon. As the detection model $\mathcal{M}$ is built only with the "*NORMAL*" class data, the task can be seen as a *one-class text classification problem*.

Conceptually, the judgment of the novelty of a text is subjective and might differ from person to person. However, there are some texts for which a majority of people are in agreement about their novelty. A good example of such a majority view of novelty are the widely spread meme pictures on social media, which contain novel interactions between objects. In this work, we restrict our research to this majority-based view (consensus view) of novelty and leave the personalized novelty view angle to future work.

This problem is difficult because of the following challenges:

1. For both types of x, fine-grained semantic reasoning is required. Although existing novelty/anomaly detection and one-class classification algorithms are applicable, they perform poorly on the task because they are coarse-grained and topic-based methods (see Sects. 6.4.3 and 6.5.3), while the proposed problem requires a contextual or semantic level of novelty detection.
2. When x is a factual text involving named entities, it requires fine-grained reasoning over (1) the relationship between the pair of entities in the textual context and (2) the background knowledge of the entities.
3. The task is a one-class classification problem. Because a one-class classification problem primarily learns from a training set containing only one class, it is generally more difficult than the traditional classification problem, which has a training set with instances from all classes.

The rest of this chapter is organized as follows: In Sect. 6.4, we assume that x is a scene description and propose the task—Semantic Novelty Detection in Scene Descriptions (SND-SD). A GAT-based model is proposed to solve this problem, which employs a graph neural network on the parsed graph representation of a text for fine-grained semantic reasoning. A novel data augmentation technique is designed to convert the one-class classification task to a supervised learning task.

In Sect. 6.5, we assume that x is a factual text involving two named entities and propose a new task—Semantic Novelty Detection in Factual Texts (SND-FT). A new model called PAT-SND is proposed to support joint reasoning over both textual context and the entities' background knowledge. A novel data augmentation technique is designed to convert the one-class classification task to a supervised learning task.

6.2   Dissimilarity and Regret

- **World** $\mathcal{W}$ **and the World Dimensionality** d'

  (1) When the input text is a scene description, the proposed task is Semantic Novelty Detection in Scene Descriptions (SND-SD). For task SND-SD, the world is the subset of scene descriptions in the physical world sampled based on some distribution. We can view the scene descriptions as a set of objects and their interactions in the 3D space that can be described with a set of verbs and other words. The world is sampled in terms of a set of verbs involved in scene descriptions. In this task, we have a large number of physical world scenes that can be described by one or more verbs. It is important to note that in this task the actual physical scenes are not observable by the agent, which is only provided with the natural language descriptions of the scenes. The novelty detection experiment is designed by giving the system semantically normal world scenes and semantically novel scenes which are annotated for evaluation. (2) When the input text is a factual text involving two entities, the task is Semantic Novelty Detection in Factual Texts (SND-FT). The world is the subset of factual texts in the real world sampled based on some distribution. We can view the factual text as two named entities and their relation to each other in the real world that are inferred from the textual context. In this task, we use Wikidata and Wikipedia to create the dataset. The world is sampled in terms of a set of predefined Wikidata relations involved in the factual text between two entities. The novelty experiment is designed by giving the system those semantically normal factual texts and semantically novel factual texts that are annotated for evaluation. In the deep neural network paradigm, the dimensionality d' is the dimension of the semantic representations of the description of a scene or fact in the form of embedding.

- **Perceptual Operators** $\mathcal{P}_t$, **Observation-Space** $\mathcal{O}$, **and its Dimensionality** d

  The input text is the description of a natural scene or a fact involving two named entities. In both cases, the temporal aspect of the perception is ignored. The only observed information by the agent is the natural language description of a natural scene or a fact. Thus, the observation-space is the input text with the dimension as the number of possible words used in the text, and possibly their associated semantic dimensions, e.g., their embeddings and semantic structures. In the deep neural network paradigm, the perceptual operators $\mathcal{P}_t$ are Deep Neural Networks (DNN) that convert text features into internal representations. The text features are created from large pretrained language models and external knowledge bases.

- **World Dissimilarity Functions** $D_{w,\mathcal{T}}$ **and Associated Threshold** δ_w

  Since the input is a natural language description of a natural scene or a fact, the world space dissimilarity is the same as the observation-space dissimilarity. Likewise, the same threshold can be applied to both spaces.

- **Observation Space Dissimilarity Functions** $D_{\mathcal{O},\mathcal{T}}$ **and Threshold** δ_O

  Since we treat the semantic novelty detection as a one-class classification problem, intuitively, we need a dissimilarity function which produces a score for the set of all normal samples and an individual sample. In our work, we use a slightly different way to achieve the functionality of dissimilarity function. We train a semantic novelty scorer, which is a DNN using all normal samples in the training data and the external knowledge. Then we produce novelty scores for all samples in the test data.

- **State Recognition Functions** $g_{y;t}(x = \mathcal{P}(w) : E_t)$

  For semantic novelty detection, a state recognition function is a deep neural network trained from normal training data for one-class classification to determine whether a test sample natural scene or a fact description is novel. For the SND-SD task, the proposed model is GAT-MA (see Sect. 6.4). For the SND-FT task, the proposed model is PAT-SND (see Sect. 6.5). Both GAT-MA and PAT-SND are semantic novelty scorers which produce semantic novelty scores for samples during testing.

- **Actions** $a_t(x)$ **Taken by Agent** A_g

  This application task does not perform any other actions except returning a prediction whether a test sample text is novel or not.

- **World Regret Function** $R_{w,\mathcal{T}}$

  Since the input is a natural language description of a natural scene or a fact, the world regret is the same as observation-space regret.

- **Observation-Space Regret Function** $R_{\mathcal{O};\mathcal{T}}$

  The observation-space regret is the error in the prediction. We measure the novelty ranking based on the novelty scores using AUC (area under the *receiver operation characteristic* (ROC) curve) or precision, recall, and F-score.

6.3   Novelty Types and Examples

In general, we have one type of novelty—semantic novelty (also called contextual novelty). However, it can be treated as two types of novelties for two tasks: SND-SD and SND-FT tasks. Both novelties are *world* novelties, and for both, the judgment of the novelty is subjective and might differ from person-to-person. In this chapter, we discuss the majority view of novelty and leave the personalized novelty view angle to future work. We briefly summarize what we have discussed in Sect. 6.1.

Scene Description Novelty. When the text is a scene description, we have scene description novelty described by novel interaction between entities/objects in the text. It is a normal scene that "A person walks a dog in the park," but it is a novel and unexpected scene that "A man is walking a chicken in the park."

Factual Text Novelty. When the text is a factual text, the factual text novelty is described by the novel interaction between two named entities. In Fig. 6.1, x_3 represents a normal fact as it is natural for an actor (Johnny Galecki) to act in the sitcom or TV show (The Big Bang Theory). However, x_4 in Fig. 6.1 depicts a novel fact with respect to the underlined entity pair because a CEO of a technology company (Elon Musk) acting in the sitcom (The Big Bang Theory) is quite surprising and novel.

6.4   Semantic Novelty Detection in Scene Descriptions

When x is a scene description, we propose the new problem of Semantic Novelty Detection in Scene Descriptions (SND-SD) to identify novel scenes from NL scene descriptions.

A new technique, called GAT-MA (*Graph Attention network with Max-Margin loss and knowledge-based contrastive data Augmentation*), is presented to solve the problem. Since our task is at the *sentence level* and *fine-grained*, we exploit Graph Attention Network (GAT) on the parsed dependency graph of each sentence, which fuses both semantic and syntactic information in the sentence for reasoning with the internal interactions of entities and actions. To enable the model to capture long-range interactions, we stack multiple layers of GATs to build a deep GAT model with multi-hop graph attention. Since SND-SD is a one-class classification problem, we also create the pseudo novel training data based on the given normal training data through contrastive data augmentation. Thus, GAT-MA is trained with the given original normal scene descriptions and the augmented pseudo novel scene descriptions (Sect. 6.4.2).

GAT-MA is evaluated using our newly created **N**ovel **S**cene **D**escription **D**etection (**NSD2**) dataset. The results show that GAT-MA outperforms a wide range of the latest novelty or anomaly detection baselines by large margins.

6.4.1   Dataset Collection and Annotation

As there is no semantic novelty detection dataset available for text, we build a new dataset, named NSD2. As our proposed task requires learning of *latent* semantic knowledge in text, such as capturing the interaction among entities and verbs, we chose three popular benchmark image caption datasets: COCO [16, 17], Flickr30k [18], and Visual Genome [19] to build our dataset. Given the merged NL caption dataset, we proceed to build our proposed NSD2 dataset as follows. We consider the captions from the NL caption dataset as normal or common scene descriptions. As our proposed GAT-MA model uses only "NORMAL" class

Table 6.1 NSD2 dataset statistics. NR (NV) denotes NORMAL (NOVEL) class and "description length" denotes the number of words

	Training	Test
Number of instances (descriptions)	202,681 (NR)	2000 (NR), 2000 (NV)
Avg. description length	11.25	11.10

data, we build our **training dataset** involving only normal scene descriptions and compile a test dataset having scene descriptions involving both **"NORMAL"** and **"NOVEL"** classes. Table 6.1 shows the summary of our NSD2 dataset statistics. The details of building the training dataset and annotating the test dataset can be found in [10].

6.4.2 Proposed GAT-MA Model

The proposed GAT-MA model consists of two main components: **(i) Knowledge-based Contrastive Data Generator (CDG)**, and **(ii) Text Semantic Novelty Scorer (SNS)**. Given a set of NL descriptions $\mathcal{X}^{tr} = \{x_1, x_2, ..., x_n\}$ of normal scenes in the training data, CDG dynamically generates *pseudo-novel descriptions* by perturbing the normal scene descriptions in $\mathcal{X}^{tr}$ utilizing the lexical knowledge-base WordNet[3] [20]. The normal descriptions in $\mathcal{X}^{tr}$ are augmented with these pseudo-novel descriptions (used as NOVEL class examples in training) to learn a SNS.

The SNS is a deep GAT model that learns to score an input text to measure its semantic novelty with respect to $\mathcal{X}^{tr}$. To capture the semantic and syntactic information in an input text x, GAT-MA parses x into a dependency graph and feeds the graph, enriched with additional word-level features, to the SNS, which is then trained to assign a higher score to a normal scene description compared to that of a novel one.

6.4.2.1 (i) Knowledge-based Contrastive Data Generator (CDG)

We propose to use the lexical knowledge-base WordNet to help generate contrastive instances to the normal scene descriptions in $\mathcal{X}^{tr}$. These contrastive instances serve as pseudo-novel data for supervised learning of the Text Semantic Novelty Scorer (SNS). WordNet contains a rich taxonomy of words and, thus, is beneficial to our semantic novelty detection task.

In our generator, a **knowledge-based misfit sampler**, $S_{misfit}(.)$, is the key component. Given a normal scene description $x \in \mathcal{X}^{tr}$, $S_{misfit}(e)$ [here, e is an entity, either a noun or a noun phrase] samples an entity e' that is semantically distant from e in the WordNet. We use Wu-Palmer Similarity [21] to measure the semantic distance between e and e'. We randomly sample e' from WordNet such that the similarity score between e and e' is less

---

[3] https://wordnet.princeton.edu/.

than 0.9 (an empirically set threshold). Next, since e' is semantically distant from e, e' is a misfit in original description x. e is replaced with e' in description x to generate a pseudo-novel description. For example, "*a man is driving a car*" describes a normal scene. It is commonsense that the subject for the verb "drive" should be a person. Anything outside of the category introduces novelty, e.g., "*a **dog** is driving a car*."

When replacing the entity e, the choice of e is also critical. For our task, we focus on three novelty aspects in a given description: (1) what actions an entity can perform, (2) what actions an entity can be applied to, and (3) how several entities interact with each other. In the interactions between entities and verbs, verbs are the core of these interactions. Thus, we only replace entities that are syntactically related to a verb to create pseudo-novel descriptions. We refer to the verb of interest in x as the *target verb*, which is used later for SNS. For details about finding and extracting entities syntactically related to the target verb, please refer to the original work [10].

Note, the novel scene description x' generated by the perturbation is contrastive to the original description x. We dynamically generate one (empirically set) pseudo-novel description for each normal description in $\mathcal{X}^{tr}$ in every training epoch.

6.4.2.2 (ii) Text Semantic Novelty Scorer (SNS)

The recent progress of employing GAT [22] on text data [23–25] has shown the advantage of explicitly combining syntactic structure (the dependency parse graph) and word-level semantics for fine-grained text analysis, such as aspect-level sentiment analysis and argument mining. Because our task is inherently a fine-grained semantic reasoning task, we build SNS based on GAT. GAT fuses the graph-structured information and node features by employing masked self-attention layers. The masked self-attention layers allow a node to attend to its neighborhood features and learn different attention weights for different neighboring nodes for graph representation learning.

Input Representation. We use a dependency parser [26] to convert an input scene description x into a dependency parse graph. For a description $x = \{w_1, w_2, ...w_n\}$, a word w_i corresponds to a node n_i in the graph. The node feature of n_i is a word embedding vector: $X_i \in \mathbb{R}^F$. F is the word embedding size. Since a description contains n words, the input node feature matrix is $X \in \mathbb{R}^{n \times F}$.

Enriching Entity Word Embeddings with Hypernym Information. We consider a noun or a noun phrase in x as an entity if it exists in the WordNet. And we refer the word(s) comprising the entity as entity word(s) and the corresponding word embedding(s) as entity word embeddings(s) onwards. Intuitively, the hypernym information of entities is beneficial to our task. Consider a normal description, "*a golden retriever is chasing a flying frisbee.*" One of the hypernym chains of the entity "*golden retriever*" in WordNet is: {golden retriever}[4] $\Rightarrow ... \Rightarrow$ {dog, domestic dog, Canis familiaris} $\Rightarrow ... \Rightarrow$ {carnivore}

---

[4] We show a synset in the format of a list of lemma names to make a synset more informative to demonstrate.

$\Rightarrow \ldots \Rightarrow$ {mammal, mammalian} $\Rightarrow \ldots$ {entity}. This hypernym chain tells us that *golden retriever* is a breed of dog. If we leverage the hypernym information, the model can not only learn that one specific breed of dog, like *"golden retriever"*, can chase a frisbee, but also generalize to other breeds of dogs as well. Additionally, this hypernym chain also contains other commonsense knowledge such as *"dogs eat meat,"* since dogs belong to the category *"carnivore."*

We perform the following three steps to incorporate hypernym features into GAT-MA:

Step-1. Candidate Entity Set Extraction. We incorporate hypernym features to entities that are syntactically related to the target verb in a description. We call these entities the candidate entities onwards. Given an input description x, this step extracts the candidate entities from x using a rule-based extractor that leverages dependency parsing and POS tagging information. Details of the method can be found in Appendix Sect. B of the original work [10]. Considering the aforementioned example, the candidate entities are *"golden retriever"* and *"frisbee"* and the target verb is *"chase."*

Step-2. Obtaining Hypernym Name Set from WordNet. Given an entity e, the Hypernym Name Set of e is the set of synset names of hypernyms of e in the WordNet. Considering the entity *"golden retriever,"* we obtain its **Hypernym Name Set** from WordNet as follows:

1. **Obtain the Synet of the Entity**. The concept of *hypernym* is defined between synsets in the WordNet. The word sense of an entity e defined in the description context corresponds to a synset in the WordNet. Ideally, a Word Sense Disambiguation (WSD) model should be employed to tag this entity with an appropriate synset. We have tried state-of-the-art WSD models, and found they do not work well with our dataset. On analysis, we found that choosing the first sense of the entity works better. Note that, according to the WordNet documentation.[5] *"Senses in WordNet are generally ordered from most to least frequently used, with the most common sense numbered 1,"* which conforms to our findings.
2. **Find the Complete Hypernym Synset Set.** With the chosen synset of the entity, we recursively collect the set of all hypernym synsets from the WordNet. For instance, given the entity *"golden retriever,"* the set of synsets in all hypernym chains originating from {golden retriever} synset to {entity} synset in the WordNet hypernym hierarchy forms the Hypernym Synset Set of *"golden retriever."*
3. **Filter Out General Hypernym Synsets.** In practice when compiling the entity hypernym information, we do not consider the whole Hypernym Synset Set for that entity because some hypernyms are too general to contribute useful knowledge for our task. Thus, we manually collect a set of synsets that are too general and remove them from the complete Hypernym Synset Set of the entity. The 24 general sysnets are given in Appendix C of the original work [10].
4. **Get the Hypernym Name Set.** An hypernym synset contains a set of lemma names. For example, given a hypernym synset—Synset('dog.n.01') of entity *"golden retriever,"*

---

[5] https://wordnet.princeton.edu/documentation/wndb5wn.

"*dog,*" "*domestic dog,*" and "*Canis familiaris*" are the lemma names. We obtain the Hypernym Name Set of an entity by collecting all lemma names from all synsets in the Hypernym Synset Set of the entity.

Step-3. Construction of the Hypernym Feature Vector. A Hypernym Feature Vector is created for each entity based on its Hypernym Name Set and is computed as the pointwise addition of all Hypernym Name Embeddings, one for each Hypernym Name in the Hypernym Name Set of the entity. We use two types of Hypernym Name Embeddings as follows:

- **GloVe-based Hypernym Name Embedding.** For a single-word hypernym name, the Hypernym Name Embedding is the corresponding GloVe word embedding [27]. For a multi-word Hypernym Name, it is computed as the average of GloVe embeddings of the words in the Hypernym Name.
- **BERT based Hypernym Name Embedding.** Since BERT produces contextual embedding for each word [28], the input of BERT should contain the context information. Given an input description, we replace the entity in the description with the Hypernym Name and feed this description into BERT. Because BERT tokenizer segments words into word pieces (subword tokens), we average the embeddings of all word pieces corresponding to this Hypernym Name to obtain the final Hypernym Name Embedding.

The Hypernym Feature Vector F^{hyper} is calculated as: $F^{hyper} = \sum_{k=1}^{M} X_k^{hyper}$, where X_k^{hyper} is the embedding of the kth Hypernym Name in the Hypernym Name Set of an entity, and M is the size of the Hypernym Name Set.

Modeling Dependency Using Deep GAT. We observe that the dependency parse graph of description x contains rich syntactic information that is beneficial to explicitly learn the interactions between entities and actions in a scene description, especially long range interactions. For a novel description like "*a monkey with a white beard and brown hair is driving a car down the street,*" the interaction among *monkey*, *drive*, and *car* makes it semantically novel. Note that, entity "*monkey*" and verb "*drive*" have a sequential word distance of 9 making it difficult for a sequential representation learning method to model the interaction. In contrast, "*monkey*" and "*drive*" are only one hop away in the dependency parse tree.

In addition, we find that for these three key words, "*drive*" is the parent of both "*monkey*" and "*car*" in the original directed dependency graph. To encourage interactions between them and allow the semantic information to flow freely in the dependency graph structure during training, we simplify the original directed dependency graph into an undirected graph. Importantly, the GAT model is trained not to attend to all neighbors of a given node equally. The attention weights to neighbors are trained to give higher weights to those nodes more useful for the task.

Fig. 6.2 Working of GAT-MA
on an input text

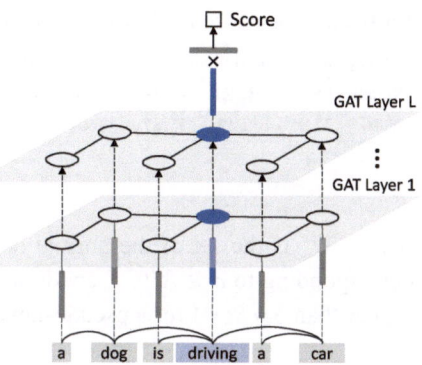

The input-output for a single GAT layer is summarized as $\boldsymbol{H}^{out} = GAT(\boldsymbol{X}, \boldsymbol{A}; \Theta)$. The input is $\boldsymbol{X} \in \mathbb{R}^{n \times F}$ and the output is $\boldsymbol{H}^{out} \in \mathbb{R}^{n \times F'}$, where n is the number of nodes, F is the node feature size, F' is GAT hidden size, and the dependency graph structure is encoded into $\boldsymbol{A} \in \mathbb{R}^{n \times n}$, which is the adjacency matrix of the graph.

In a single GAT layer, a word or an entity in a graph only attends over the local information from 1-hop neighbors. To enable the model to capture long-range interactions between entities and actions, we stack L layers to make a *deep* model, which allows information from L-hops away to propagate into this word.

As illustrated in Fig. 6.2, the stacking architecture is represented as $\boldsymbol{H}^{l+1} = GAT(\boldsymbol{H}^l, \boldsymbol{A}; \Theta^l)$, $l \geq 0$, $\boldsymbol{H}^0 = \boldsymbol{X}\boldsymbol{W}_0 + \boldsymbol{b}_0$. The output of the GAT layer l, $\boldsymbol{H}^l_{out} = GAT(\boldsymbol{H}^l, \boldsymbol{A}; \Theta^l)$, is the input for layer $(l+1)$, denoted by $\boldsymbol{H}^{l+1}$. $\boldsymbol{H}^0$ is the initial input. $\boldsymbol{W}_0 \in \mathbb{R}^{F \times F'}$ and $\boldsymbol{b}_0$ are the projection matrix and bias vector. For a L layer GAT-MA model, the output of the final layer is $\boldsymbol{H}^L_{out} \in \mathbb{R}^{n \times F'}$.

For our task, we are concerned with interactions of verbs and entities. As mentioned in the discussion of CDG, when perturbing the normal descriptions, we only replace the entities that are syntactically related to a verb in the dependency graph. This verb is our target verb. Any novelty introduced in the description due to the replacement is related to this verb. If a description contains multiple verbs, the target verb of an entity is the one which is close to it along the dependency parse graph.

We use a mask layer $\boldsymbol{m}$ to fetch the output embedding for this target verb v_i from GAT: $\boldsymbol{h}_{v_i} = \boldsymbol{m}\boldsymbol{H}^L_{out}$, where $\boldsymbol{m} \in \mathbb{R}^{1 \times n}$ is a one-hot vector indicating the position of the target verb. Next, we use a feed-forward layer to project $\boldsymbol{h}_{v_i}$ into a semantic novelty score. We denote the score function of SNS by $SNS(x)$ for the input description x.

Training. GAT-MA is trained end-to-end by minimizing a max-margin ranking objective, as given below,

$$\mathcal{L} = \sum_{x \in \mathcal{X}^{tr}} \sum_{x' \in \mathcal{X}'} max\{SNS(x') - SNS(x) + 1, 0\}, \tag{6.1}$$

Table 6.2 Comparison of baselines and our proposed model (based on AUC score)

Language model based model				General One-class classifier							Proposed
Ngram	LSTM	BERT	GPT-2	OCSVM	iForest	VAE	DSVDD	ICS	OCGAN	HRN	GAT-MA
76.76	77.95	82.13	77.87	68.07	50.55	51.43	54.89	56.15	50.80	56.83	**89.22**

where $\mathcal{X}^{tr}$ is the set of the normal descriptions, $x' \in \mathcal{X}'$ is the pseudo-novel description corresponding to $d \in \mathcal{X}^{tr}$. $\mathcal{L}$ encourages the score $SNS(x)$ of normal description x to be higher than $SNS(x')$ for a pseudo-novel description x'.

6.4.3 Experiments

For dataset details, please refer to Sect. 6.4.1. The Appendix of the original work [10] has additional information about the data and model implementation details.[6]

Baselines. We compare GAT-MA with three categories of baselines: (1) four language model-based novelty detection models, (2) seven one-class classification models, and (3) other models based on different text encoders and loss functions. All the results in this section are the average of five runs with different seeds. The results are statistically significant with $p < 0.001$.

A trained Language Model (LM) can be intuitively used as a novelty detection model due to the following reasons: (1) When training an LM on normal scene descriptions, the model minimizes the perplexity of the training data by maximizing the likelihood of each word appearing in its context. In this way, it indirectly learns the semantic meaning of words and sentences. (2) Each LM trained on normal descriptions can output the probability of each word in a description appearing in its context. Thus, a sentence probability can be calculated from the list of word probabilities. We have tried various ways of calculating the sentence score from the word probability list, such as arithmetic mean, geometric mean, harmonic mean, and multiplication of all word probabilities and found harmonic mean to be the best choice. We use **N-gram**, the bag of words LM, $N \in \{1, 2, 3, 4, 5\}$ ($N = 1$ gives the best result), **LSTM** [29], **BERT** [30], **GPT-2** [31] as our LM baselines. The results are listed in Table 6.2.

For general one-class classification models, most of them only work on images. We modified the related components of the models to make them suitable for text data. More details regarding model modification and parameter setting are provided in Appendix F of the original paper [10]. The following seven baselines are compared: (1) **DSVDD** (Deep SVDD) [32]: a one-class classifier, which is the deep learning version of SVDD. (2) **ICS** [33]: a one-class classification method trained on one class of training data that is split into two

---

[6] The code and the annotated dataset are released at: https://github.com/NianzuMa/semantic-novelty-detection-in-natural-language-descriptions.

subsets: typical and atypical. (3) **OCGAN**[34]: a one-class anomaly detection method based on GAN. (4) **VAE** [35]: the variational auto-encoder. (5) **OCSVM** [36]: the classic SVM method for one-class classification. (6) **iForest** [37]: a classic ensemble method based on random unsupervised trees. (7) **HRN** [38]: the recent model based on holistic regularization. We could not compare with another recent baseline CSI [39] as it is based on various image transformations. We do not compare with out-of-distribution (OOD) detection methods as they require multiple classes to learn.

Experiment settings. In general, we conducted experiments using various word and sentence embeddings, such as GloVe[7] [27], BERT[8] [30], and InferSent [40, 41]. We only show the best results in Table 6.2. The detailed hyper-parameter settings for GAT-MA and baseline models are included in Appendice E and F of the original work [10].

Evaluation Metrics. Following the existing novelty/anomaly detection literature [4, 42], we produced only a score function and ignored the binary decision problem and, thus, used AUC (Area Under the ROC curve) as the evaluation metric. All compared models are trained with only normal scene descriptions.

6.4.3.1 Results and Analysis

Baseline Comparison. Table 6.2 shows the predictive performance comparison of the base-lines and our proposed model GAT-MA, which is based on BERT embedding and enhanced with hypernym embedding features. From Table 6.2, we conclude the following:

(1) All general one-class classifiers perform poorly on our task. Even the reported state-of-the-art model HRN gives an AUC score of only 56.83. We have tried various ways to produce the description embedding as the input feature for these models, such as (a) averaging all words' GloVe embeddings, (b) feeding the description into BERT and using the first token [CLS]'s embedding as the sentence embedding, (c) feeding the description into BERT and averaging all output tokens' embeddings as the sentence embedding, and (d) feeding the description into the pre-trained sentence embedding extractor InferSent to produce the sentence embedding. However, none of these options give good performances. These one-class classifiers perform well on image data because images of a given class, e.g., the MNIST dataset, contains images with very similar latent representations. Thus, auto-encoder and GAN-based models can learn latent representations for all instances in an image class very close to each other in the latent space. In contrast, our normal scene descriptions have many topics and it is hard for them to learn latent representations that are close to each other in the latent space.

(2) Language model-based methods are in general better than one-class classifiers because they, in some sense, do not try to learn a latent representation, but exploit the sequential

---

[7] We use glove.840B.300d in our experiments.

[8] We use the BERT model "bert-base-uncased" as text encoder. We expect that using larger transformer embeddings leads to better results. But due to the limitation of our computing resources, we have to use this base BERT model.

Fig. 6.3 Effects of the number
of layers in GAT-MA$_{vanilla}$

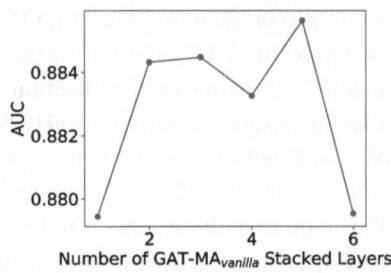

and semantic information of the input text to produce word probabilities. Thus, they are comparatively more effective in fine-grained semantic novelty detection. However, they still perform much worse than GAT-MA as they mainly learn the word distribution in the normal description data but do not explicitly capture the interaction of entities and verbs.

In summary, GAT-MA outperforms all baselines by large margins and is more effective for our proposed task. Below we discuss ablation and additional experiments.

Effects of Word Embedding and Hypernyms. In Table 6.4, GAT-MA$_{vanilla}$ is our proposed model using BERT embedding without being enhanced with the hypernum features. GAT-MA$_{GloVe}$ is our proposed model using GloVe embedding without being enhanced with hypernym features. Comparing the results of GAT-MA$_{GloVe}$ and GAT-MA$_{vanilla}$, we can see that BERT embedding contains richer semantic knowledge which is more beneficial to our task compared to using GloVe embedding. It is also interesting to see that when GAT-MA$_{vanilla}$ is enhanced with hypernym embedding feature (noted as GAT-MA), it improves the AUC score from 88.12 to 89.22, which means that hypernym features can help our model generalize better.

Effects of Model Depth. From Fig. 6.3, we see that increasing the number of stacked layers from one to five improves the performance of GAT-MA$_{vanilla}$. When the number of stacked layers is higher than five, the performance drops. This is because most of the interactions happen between entities and actions that are near each other in the dependency parse graph. Stacking five layers is enough and more stacked layers will not help but hurt the performance (Fig. 6.3).

Effects of Using Max-margin Ranking Loss. Table 6.5 compares fine-tuned BERT and GAT-MA variants in terms of the use of loss functions in model training. Here, $[\cdot]_{CE}$ denotes the model using the cross entropy loss for training and $[\cdot]_{MM}$ denotes the model using the max-margin loss as proposed in Eq. 6.1. From Table 6.5, we see that CE variants are weaker than MM variants for both BERT and GAT-MA. Both GAT-MA$_{CE}$ and GAT-MA$_{MM}$ use BERT embeddings without the hypernym feature.

Effects of Using Dependency Parse Structure. Table 6.5 shows that BERT$_{MM}$, not directly using any syntactic features, easily fails on examples dissimilar to training data in terms of word distribution. However, GAT-MA$_{MM}$ performs better by explicitly modeling the dependency parse structure. This means that modeling dependency parse structure is

Table 6.3 Some descriptions predicted wrongly by BERT$_{MM}$ but correctly by GAT-MA$_{MM}$

	Text	Label
1	A monkey with glasses is cooking food on a stovetop in a kitchen	Novel
2	A couple of seal dogs carry their surfboard across the beach	Novel
3	A giant panda in a white smock prepares to cut the hair of an older balding gentleman in front of a case holding several hair supplies	Novel
4	An adult is walking on the sidewalk in St. Louis	Normal
5	A guy eats food on a table in front of a food shop on the street while a passerby walks by	Normal
6	A group of people standing around are drinking some vermouth	Normal

Table 6.4 Effect of using embedding type and hypernym feature based on AUC scores

GAT-MA$_{GloVe}$	GAT-MA$_{vanilla}$	GAT-MA
84.42	88.12	89.22

Table 6.5 Comparison of BERT and GAT-MA variants based on cross-entropy (CE) and max-margin (MM) loss function based on AUC scores

BERT$_{CE}$	BERT$_{MM}$	GAT-MA$_{CE}$	GAT-MA$_{MM}$
82.09	87.41	83.80	88.12

beneficial to capturing the interactions between entities and actions. Some descriptions predicted incorrectly by BERT$_{MM}$ but correctly by GAT-MA$_{MM}$ are shown in Table 6.3.

Error Analysis. The AUC score in (.) for each verb is as follows: pull (0.99), carry (0.99), push (0.99), drive (0.97), travel (0.97), hit (0.95), throw (0.95), kick (0.94), climb (0.94), look (0.93), build (0.93), cook (0.92), walk (0.92), ride (0.87), fly (0.84), cut (0.82), swim (0.81), jump (0.73), drink (0.73), and eat (0.69).

We carried out error analysis on our test data and found that the errors are mainly due to the following factors. The first factor is the pretrained word embedding quality. The quality of the word embedding is critical for GAT-MA to effectively do reasoning. GAT-MA makes mistakes when the pretrained word embedding is not of good quality. For example, the *"talapoin"* in *"the talapoin at the zoo is leaning down to drink some water."* The second factor is the limitation of knowledge acquired by GAT-MA during training. GAT-MA relies on the taxonomy information in WordNet to generate contrastive novel descriptions during training. However, sometimes the reasoning of the novel description requires more complex world knowledge. For example, "two kids are sitting in the bar drinking spirit" is novel and requires knowledge that kids are not old enough to drink any alcohol. Another example "A dog is eating onions on the ground" is novel and requires the world knowledge that onions are poisonous to dogs.[9]

6.5   Semantic Novelty Detection in Factual Texts

This subsection proposes and studies a new problem, called *Semantic Novelty Detection in Factual Texts* (SND-FT), that involve named entities. SND-FT is significantly different from the previous *Semantic Novelty Detection in Scene Descriptions* (SND-SD) problem because solving the new problem requires fine-grained reasoning over both (1) the *relationship* between the pair of entities in the textual context and (2) the *background knowledge* of the entities. For example, considering x_3 in Fig. 6.1, we first need to detect that the entity pair ("*Elon Musk*," "*The Big Bang Theory*") in x_3 has the "*cast-member*" relation and then, leverage the interaction of the relation with the background knowledge of the entities, i.e., "*Elon Musk*" is a *tech entrepreneur* and "*The Big Bang Theory*" is a *TV show*, to infer the semantic novelty (because a tech entrepreneur does not normally act in a TV show). For this work, an external Knowledge Repository (KR) is needed. We use WikiData [43] to extract a named entity's background knowledge, i.e., a list of property-value pairs. For example, *Elon Musk*'s background knowledge contains property-value pairs: [(a) (occupation, *entrepreneur*), (b) (gender, *male*), (c) (field of work, *tech entrepreneurship*) ....]. However, not all property-value pairs are useful for inference, e.g., only (a) and (c) are useful for x_3 in Fig. 6.1. Thus, a solution for automatic selection of the *useful* property-value pairs is needed (see Sect. 6.5.2).

Based on our analysis of the SND-FT task above, we extend the general problem definition of fine-grained semantic reasoning in Sect. 6.1 and present a specific problem definition for the SND-FT task below:

SND-FT Problem Definition: Given (1) a set of training factual text $\mathcal{X}_{tr} = \{x_1, x_2, ...x_n\}$ with each $x_i \in \mathcal{X}_{tr}$ labeled as normal (*NORMAL* class) with respect to a pair of entities (e_1^i, e_2^i) appeared in x_i, and (2) a knowledge base (KB) $\mathcal{K}$ containing the background knowledge (property-value pairs) of a set of entities that is a superset of the entities appeared in $\mathcal{X}_{tr}$, our goal is to build a model $\mathcal{F}$ to score the semantic novelty of a test factual text x' having a pair of entities (e_1', e_2') with respect to $\mathcal{X}_{tr}$, $\mathcal{K}$, and pair (e_1', e_2'), i.e., classifying x' into one of the two classes {*NORMAL, NOVEL*}. As $\mathcal{F}$ is built with only the "*NORMAL*" data, the task is again a *one-class classification problem*.

This task is different from the SND-SD task [10] in two main aspects: (1) The new task requires semantic reasoning over named entities which do not have sufficient semantic information in their textual (or surface) form. Rich background knowledge of the entities is needed to detect novelty. The SND-SD task does not require any such entity background knowledge. (2) The SND-SD task does semantic reasoning for relations (between objects), based on a fixed/closed set of verbs. However, for the new problem, the relations between

---

[9] https://en.wikipedia.org/wiki/Dog_health.

entities may be expressed in any surface forms and/or even implicitly, e.g., the relation "*cast-member*" between the underlined entities is expressed implicitly in x_3. Thus, the GAT-MA model for the SND-SD task cannot handle such cases.

To solve the new problem, a new model is proposed, called PAT-SND (*Property ATtention network for Semantic Novelty Detection*). PAT-SND first employs an existing relation classification technique to identify the relation between the entity pair. The identified relation is then used in a novel *relation-aware* **P**roperty **AT**tention Network (PAT) module that leverages the attention mechanism to select the useful background knowledge from the KB $\mathcal{K}$ to perform semantic reasoning for novelty detection. PAT-SND is evaluated using a newly created NFTD (Novel Factual Text Detection) dataset since there is no available data that is suitable for our evaluation. We leverage a distant supervision technique [44] with Wikipedia[10] as the corpus and Wikidata as the KR to build a large training dataset. Evaluation results show that PAT-SND outperforms ten of the latest novelty detection baselines by very large margins.

In the following sub-sections, we first discuss how we collect and annotate the NFTD dataset in Sect. 6.5.1. Next, we present the details of the proposed PAT-SND model in Sect. 6.5.2. In Sect. 6.5.3, we evaluate the PAT-SND model using the NFTD dataset.

6.5.1 Dataset Collection and Annotation

To build a large factual text dataset annotated with named entities, we leverage the distant supervision technique in [44]. We create our training and test datasets using Wikipedia as the corpus and Wikidata [43] as the external Knowledge Repository (KR). The training data is sampled directly from the dataset created using the distant supervision technique. The test dataset are annotated by crowdsourcing workers. The details of the creation of the training and test sets are given in Sect. 3 of the original work [11]. Table 6.6 shows the NFTD dataset statistics.

Building Entity Background KB. We use the Knowledge Repository (KR), Wikidata, to build the entity background KB $\mathcal{K}$. KR is represented as: $KR = (\mathcal{E}, \mathcal{R}, \mathcal{T})$, where $\mathcal{E}$ denotes a set of entities, $\mathcal{R}$ is a set of relations/edges, and $\mathcal{T} \subseteq \mathcal{E} \times \mathcal{R} \times \mathcal{E}$ is the set of all triples. For each entity e in $\mathcal{E}$, we obtain the list of property-value pairs as e's background

Table 6.6 NFTD dataset statistics. NR (NV) denotes the NORMAL (NOVEL) class and "text length" is the number of words

	Training	Test
# instances (factual text)	251,619 (NR)	1000 (NR), 1000 (NV)
Avg. text length	41.35	26.02

---

[10] https://en.wikipedia.org/wiki/Main_Page.

knowledge to build $\mathcal{K}$ as follows: We first collect all triples from KR involving e and then extract the relation and the other entity from each triple to form a property-value pair with the relation as a property and the other entity as the value of the property. For example, considering $e = $ "*Elon Mask*" and a triple ("*Elon Mask*," "*occupation*," and "*entrepreneur*") in KR, the extracted property-value pair for e would be (`occupation`, *entrepreneur*).

Let $\mathcal{P}$ be the complete property set in the background KB $\mathcal{K}$. We assume that each e_i in the training data is in $\mathcal{K}$. However, e_i in the test data can be a new entity, i.e., it does not appear in the training data, as long as the background knowledge of the entity is available to our model (where, the property-value pairs are either retrieved from the KR or provided by the human annotators during the test data annotation process. The properties provided by the human annotators belong to $\mathcal{P}$. The values provided by the human annotators can be any string value).

6.5.2 Proposed Approach

The proposed PAT-SND model works in two steps: (1) *Entity Relation Classification* and (2) *Triple Semantic Novelty Scoring* (SNS). Given a factual text x containing a pair of entities (e_1, e_2), PAT-SND first identifies the relation $\hat{r}$ between (e_1, e_2) in x in step one. Next, the background knowledge of the entities e_1 and e_2 retrieved from the KB $\mathcal{K}$ together with the predicted relation $\hat{r}$ are fed to the SNS module to score the semantic novelty of x with respect to (e_1, e_2) and $\mathcal{K}$ in step two. As our training data $\mathcal{X}_{tr}$ consists of only NORMAL class examples, it is not possible to train SNS solely with $\mathcal{X}_{tr}$. Thus, we propose a *KB-based Contrastive Data Generator* (CDG) to generate pseudo-novel examples. The SNS module is then trained with both NORMAL class examples in $\mathcal{X}_{tr}$ as well as the generated pseudo-novel examples in a supervised learning manner.

Step 1: Entity Pair Relation Classification. Given a factual text x having entity pair (e_1, e_2), we build a model to identify the relation $\hat{r}$ between (e_1 and e_2) in x. For this purpose, we utilize a BERT-based Relation Classification model [45] that incorporates entity position information into a pre-trained language model for relation classification. Next, we combine the identified relation $\hat{r}$ with the entity pair to produce a triple $(e_1, \hat{r}, e_2)$ which serves as input to the SNS.

During the training process, the relation classification model is trained using $\mathcal{X}_{tr}$, where each $x_i \in \mathcal{X}_{tr}$ is labelled with *true relation label* r between the entity pair through a distant supervision technique.

Step 2: Triple Semantic Novelty Scoring (SNS). Let $B_1 = \{(p_i^1, v_i^1)|1 \le i \le l\}$ and $B_2 = \{(p_i^2, v_i^2)|1 \le i \le m\}$ be the background knowledge obtained for e_1 and e_2, respectively, from KB $\mathcal{K}$ (see Sect. 6.5.1). The SNS module utilizes B_1, B_2, and relation $\hat{r}$ as inputs to score the novelty of the input text x. In this process, SNS employs a relation-aware attention mechanism over B_1 and B_2 to select the useful knowledge, which is motivated as follows.

e_1: The Big Bang Theory		e_2: Elon Musk	
Property	*Value*	*Property*	*Value*
instance of	television series	**instance of**	human
start time	24 September 2007	**gender**	male
end time	16 May 2019	**occupation**	entrepreneur
audio system	Dolby Digital	**field of work**	tech entrepreneurship
...		...	

Fig. 6.4 Illustration of two entities' property and value pairs in the KB $\mathcal{K}$. The properties marked in red are useful or important for detecting semantic novelty of example x_3 in 6.1

Leveraging all property-value pairs in B_1 and B_2 may not be helpful to detect the novelty of the text x. For example, as shown in Fig. 6.4, considering the entity "*Elon Mask*," the property-value pair (`occupation`, *entrepreneur*) is useful to score the novelty of x_3 in Fig. 6.1, whereas (`gender`, *male*) is not useful at all. Thus, the model needs to have the ability to focus on the important information and filter out noises in B_1 and B_2. This knowledge selection process is relation dependent, as for different relations, different property-value pairs could be useful for novelty detection.

To enable automated knowledge selection, SNS is built using a key component called **Property Attention Network (PAT)** that utilizes the semantics of the relation $\hat{r}$ to attend over B_1 and B_2 for inference. As the attention mechanism needs to be relation-specific, we build one PAT module for each relation. So, for detecting novelty of a test text x', SNS fires the PAT learned for relation $\hat{r}$, identified from x' using the Relation Classifier.

Property Attention Network (PAT). PAT takes a list of property-value pairs, $\{(p_i, v_i)|1 \leq i \leq N\}$ and a relation r as input and outputs a weighted value vector $\boldsymbol{h}^{out}$ to be used for inference. p_i and the corresponding v_i are fed to PAT as feature vectors $\boldsymbol{p}_i, \boldsymbol{v}_i$, respectively, together with r (to invoke the relation-specific module). We employ BERT [30] to learn the embedding representation of p_i, v_i and use them as the corresponding feature vectors. For example, the property "*instance of*" is encoded as $\langle$[CLS], instance, of, [SEP]$\rangle$ using WordPiece Tokenizer and fed into BERT and the embedding corresponding to token [CLS] in the output layer of BERT is used as the feature vector of the property.

In PAT, the $\{\boldsymbol{p}_i\}_{i=1}^N$ are fed one by one through a relation-specific linear layer, and a *relu* non-linearity function with a SoftMax function are used to obtain the attention weights $\{\alpha_{ir}\}_{i=1}^N$ over $\{\boldsymbol{p}_i\}_{i=1}^N$ with respect to r. Next, the weights are used to weigh the corresponding $\{\boldsymbol{v}_i\}_{i=1}^N$ to obtain $\boldsymbol{h}^{out}$. The processing for a given r is summarized,

$$g_{ir}^k = relu(\boldsymbol{p}_i \, \boldsymbol{W}_r^k + \boldsymbol{b}_r^k)$$

$$\alpha_{ir}^k = \frac{exp(g_{ir}^k)}{\sum_{i=1}^N exp(g_{ir}^k)} \tag{6.2}$$

$$\boldsymbol{h}^{out} = \sum_{i=1}^N \left(\frac{1}{K} \sum_{k=1}^K \alpha_{ir}^k \right) \boldsymbol{v}_i,$$

where K is the total number of attention heads and $\boldsymbol{W}_r^k$, $\boldsymbol{b}_r^k$ are relation-specific weight and bias for the kth attention head. α_{ir}^k is the kth attention weight between r and p_i. Overall, the processing of inputs in PAT is denoted as $h^{out} = PAT(\boldsymbol{P}, \boldsymbol{V}, r; \Theta_r)$, where $\boldsymbol{P} = [\boldsymbol{p}_1, \boldsymbol{p}_2, ..., \boldsymbol{p}_N] \in \mathbb{R}^{N \times F}$ is the property matrix, $\boldsymbol{V} = [\boldsymbol{v}_1, \boldsymbol{v}_2, ..., \boldsymbol{v}_N] \in \mathbb{R}^{N \times F}$ is the value matrix, and Θ_r is the trainable parameters for relation r.

Triple Novelty Scoring. Given the inputs B_1, B_2, and relation $\hat{r}$, we obtain the property and value matrices $\boldsymbol{P}_1$, $\boldsymbol{V}_1$ from B_1 and $\boldsymbol{P}_2$, $\boldsymbol{V}_2$ from B_2 and feed them to PAT for relation $\hat{r}$ as follows,

$$h_1^{out} = PAT(\boldsymbol{P}_1, \boldsymbol{V}_1, \hat{r}; \Theta_{\hat{r}})$$
$$h_2^{out} = PAT(\boldsymbol{P}_2, \boldsymbol{V}_2, \hat{r}; \Theta_{\hat{r}}) \tag{6.3}$$
$$h_{\hat{r}}^{out} = [h_1^{out}; h_2^{out}].$$

Next, a relation-specific feed-forward layer is used to project $h_{\hat{r}}^{out}$ into a semantic novelty score as $SNS(\hat{\tau}) = (h_{\hat{r}}^{out} \, \boldsymbol{W}_{\hat{r}} + \boldsymbol{b}_{\hat{r}})$, where $\hat{\tau}$ denotes the triple $(e_1, \hat{r}, e_2)$. Following the existing one-class classification literature [4, 42], we do not use a threshold to further produce a classification label, instead use $SNS(\hat{\tau})$ directly in our experiments (Sect. 6.5.3).

Training. Let $\mathcal{T}_{tr}$ be the set of all triples (labelled as NORMAL class) extracted from the examples in $\mathcal{X}_{tr}$. To train SNS, we use KB $\mathcal{K}$ to help generate contrastive examples (triples) by corrupting the triples in $\mathcal{T}_{tr}$, as discussed below. These contrastive examples serve as the pseudo-novel data and enable the supervised learning of the SNS.

KB-based Contrastive Data Generator. Given a triple $\tau_i \in \mathcal{T}_{tr}$, the generator $G_{contrastive}(\tau_i)$ randomly samples an entity e' from KB $\mathcal{K}$ to replace either e_1 or e_2 in τ_i. After corruption, τ_i' is formed from τ_i, where $\tau_i' = (e', r, e_2)$ or $\tau_i' = (e_1, r, e')$. For example, given $\tau_1 = (\text{The Big Bang Theory, cast-member, Johnny Galecki})$ as a NORMAL triple in $\mathcal{T}_{tr}$, a pseudo-novel triple generated by $G_{contrastive}(\tau_1)$ would be $\tau_1' = (\text{The Big Bang Theory, cast-member, } \underline{\text{Warren Buffett}})$. During the training of SNS, we dynamically generate one pseudo-novel triple for each NORMAL triple in $\mathcal{T}_{tr}$ in every training epoch.

Learning. PAT-SND is trained end-to-end by minimizing a max-margin ranking objective as,

$$\mathcal{L} = \sum_{\tau \in \mathcal{T}_{tr}} \sum_{\tau' \in \mathcal{T}_{tr}'} max\{SNS(\tau') - SNS(\tau) + 1, 0\}, \tag{6.4}$$

where $\mathcal{T}_{tr}'$ is the set of pseudo-novel triples generated from $\mathcal{T}_{tr}$. $\mathcal{L}$ encourages the score $SNS(\tau)$ of the NORMAL triple τ to be higher than $SNS(\tau')$ of a pseudo-novel triple τ'.

6.5.3   Experiments

The details of the dataset annotation and statistics have been discussed in Sect. 6.5.1. All the results reported in this section are the averages of five runs with different random seeds. The code and the dataset have been released.[11]

Evaluation Metrics. Since our task is a one-class classification task, we again follow the existing one-class classification literature [4, 42] and use AUC (Area Under the ROC curve) as the evaluation metric.

Baselines. Since the proposed task is new, we are not aware of any existing model that can be directly applied to the task. We thus convert two types of existing methods for the task: (i) **language models (LMs)** and (ii) **traditional and deep learning-based one-class classifiers**. Note that the GAT-MA model in the previous section cannot be used here as a baseline because the model needs verbs expressed explicitly in the text for novelty scoring. However, in our case, the relation in the factual text may be implicitly expressed in various surface forms, which makes GAT-MA inapplicable to our task. Furthermore, GAT-MA cannot use the background knowledge.

(i) **LM-based SNS.** We train LMs on our training text data, which are all normal factual text. When the LMs are trained to minimize the perplexity of text, it maximizes the probability of the words appearing in the text context. The trained models thus capture the semantic meaning of words and the text. If something unexpected happens in the context, the model can detect the novelty. The trained language models are used first to output the probability of each word in the text. We then calculate the sentence probability based on these word probability scores. Following [10], we use (a) arithmetic mean, (b) geometric mean, (c) harmonic mean, and (d) multiplication of all word probabilities. We find that harmonic mean gives the best detection results. Among language models, we adopt **N-gram**, the bag of words LM, $N \in \{1, 2, 3, 4, 5\}$ ($N = 5$ gives the best result), **BERT** [30], **GPT-2** [31] as our LM-based SNS and show the results in Table 6.7.

(ii) **One-class Classifier-based SNS.** One-class classification methods aim to build classifiers in the setting that "the data observed during training is from a single positive class" [46]. There is a considerable amount of research that has been done in computer vision, machine learning, and biometrics communities. Most of them are designed for image data. We convert the models to SNSs by modifying the feature encoder parts of the models. Here are the classical statistical and recent deep learning-based one-class classifiers:

(1) **OCSVM** [36]: the classic one-class SVM classifier. (2) **iForest** [37]: an ensemble method using random unsupervised trees. (3) **VAE** [35]: a variational auto-encoder used as one-class classifier. (4) **OCGAN** [34]: a popular one-class novelty detection model based on GAN. (5) **DSVDD** (Deep SVDD) [32]: a deep learning implementation of the one-class classifier SVDD [47]. (6) **ICS** [33]: a one-class classifier trained using the training data split into two parts: typical and atypical. (7) **HRN** [38]: a recent model based on a holistic regu-

---

[11] The Github repository for released code and the annotated data: https://github.com/NianzuMa/PAT-SND.

Table 6.7 Comparison of baselines and our proposed model (based on AUC score). Each result in the table is the average of five runs with different seeds ($\pm$ standard deviation)

Language model based model			General One-class classifier							Proposed
Ngram	BERT	GPT-2	OCSVM	iForest	VAE	DSVDD	ICS	OCGAN	HRN	**PAT-SND**
50.02±0.0	60.12±0.0	58.13±0.0	50.63±0.0	44.16±1.3	47.94±0.3	51.00±0.5	53.98±0.5	52.10±0.0	55.53±1.3	**90.37±0.5**

larization method. We do not compare with other models that require image transformation such as CSI [39]. Out-of-distribution (OOD) detection methods are not applicable to our task since they typically need multiple classes to train the model. The details of experiment settings are provided in Appendix B of the original work [11].

Model Comparison and Discussion. We show the results of all baselines and our proposed model PAT-SND in Table 6.7. Here are the conclusions we can drawn from the results:

(1) All LM-based SNSs perform poorly on the proposed factual text novelty detection task because although they implicitly learn the syntactic and semantic information of the text, they cannot explicitly do semantic reasoning. The information in text alone is not enough to distinguish normal and novel factual text. Our task needs the background information (property-value pairs) of named entities to perform semantic reasoning and detect novelty. The language models dealing with sequential data can hardly incorporate background knowledge of named entities during training.

(2) All one-class classifier-based SNSs also perform poorly on our task. To employ the one-class classifiers, we first extract the text embedding using a text encoder and then use the embedding to learn the classifier. The text encoder parameters are frozen during the classifier training. The text embedding is computed by averaging the token embeddings obtained from the last layer of BERT (used as text encoder in our baselines). However, none of these methods are able to incorporate the background knowledge of the named entities into the embedding. Thus, they perform poorly on our task.

For the proposed method, the macro F1 score of relation classification is 95.12%. PAT-SND's novelty detection AUC score is 90.37, which is better than the AUC score of all baselines by large margins. We believe the reasons are: (1) the proposed model exploits the background knowledge of the two named entities to do semantic reasoning, which is a necessity for our task and (2) the contrastive data augmentation converts our task to a supervised learning task, enabling our model to be trained to select important relation-specific properties and values to do effective semantic reasoning.

6.6   Conclusions

In this chapter, we studied the issue of contextual novelty in the context of natural language processing. Two new problems and their solutions have been proposed and evaluated. The first problem is to detect novel scene descriptions. The proposed method is called GAT-MA based on a graph attention network that exploits parsing and data augmentation to solve the problem. As there was no existing evaluation dataset for the proposed task, an evaluation dataset, named NSD2, was created. Experimental comparisons with a wide range of baselines showed that GAT-MA outperformed them by very large margins. The second problem is to detect novel sentences that involve named entities. A novel attention-based network PAT-SND was proposed to solve the problem. A new evaluation dataset NFTD was also created and released as a benchmark for the NLP community. Experimental results demonstrated the effectiveness of PAT-SND. The original paper [11] also showed that the attention results can be used to characterize the discovered novelty.

References

1. Grubbs FE (1969) Procedures for detecting outlying observations in samples. Technometrics 11(1):1–21
2. Barnett V, Lewis T (1994) Outliers in statistical data. Wiley series in probability and statistics. Wiley
3. Chalapathy R, Menon AK, Chawla S (2018) Anomaly detection using one-class neural networks. arXiv:1802.06360
4. Pang G, Shen C, Cao L, Van Den Hengel A (2021) Deep learning for anomaly detection: a review. ACM Comput Surv (CSUR) 54(2):1–38
5. Fei G, Liu B (2016) Breaking the closed world assumption in text classification. In: Proceedings of the 2016 conference of the North American chapter of the association for computational linguistics: human language technologies. San Diego, California, pp 506–514
6. Shu L, Xu H, Liu B (2017) DOC: Deep open classification of text documents. In: Proceedings of the 2017 conference on empirical methods in natural language processing, Copenhagen, Denmark, pp 2911–2916
7. Xu H, Liu B, Shu L, Yu PS (2019) Open-world learning and application to product classification. In: Liu L, White RW, Mantrach A, Silvestri F, McAuley JJ, Baeza-Yates R, Zia L (eds) The world wide web conference, WWW 2019, San Francisco, CA, USA. ACM, pp 3413–3419. Accessed from 13–17 May 2019
8. Lin T-E, Xu H (2019) Deep unknown intent detection with margin loss. In: Proceedings of the 57th annual meeting of the association for computational linguistics, Florence, Italy, pp 5491–5496
9. Zheng Y, Chen G, Huang M (2020) Out-of-domain detection for natural language understanding in dialog systems. IEEE/ACM Trans Audio, Speech, Lang Process 1198–1209
10. Ma N, Politowicz A, Mazumder S, Chen J, Liu B, Robertson E, Grigsby S (2021) Semantic novelty detection in natural language descriptions. In: Proceedings of the 2021 conference on empirical methods in natural language processing, pp 866–882
11. Ma N, Mazumder S, Politowicz A, Liu B, Robertson E, Grigsby S (2022) Semantic novelty detection and characterization in factual text involving named entities. In: Proceedings of the 2022 conference on empirical methods in natural language processing, pp 9225–9252
12. O'Brien HL, Toms EG (2008) What is user engagement? a conceptual framework for defining user engagement with technology. J Am Soc Inf Sci Technol 59(6):938–955
13. O'Brien HL, Toms EG (2010) The development and evaluation of a survey to measure user engagement. J Am Soc Inf Sci Technol 61(1):50–69

14. Hassenzahl M, Tractinsky N (2006) User experience-a research agenda. Behav Inf Technol 25(2):91–97
15. Attfield S, Kazai G, Lalmas M, Piwowarski B (2011) Towards a science of user engagement (position paper). In: WSDM workshop on user modelling for Web applications, pp 9–12
16. Chen X, Fang H, Lin T-Y, Vedantam R, Gupta S, Dollár P, Zitnick CL (2015) Microsoft coco captions: Data collection and evaluation server. arXiv:1504.00325
17. Lin T-Y, Maire M, Belongie S, Hays J, Perona P, Ramanan D, Dollár P, Zitnick CL (2014) Microsoft coco: common objects in context. In: ECCV, pp 740–755
18. Plummer BA, Wang L, Cervantes CM, Caicedo JC, Hockenmaier J, Lazebnik S (2015) Flickr30k entities: Collecting region-to-phrase correspondences for richer image-to-sentence models. In: 2015 IEEE international conference on computer vision, ICCV 2015, Santiago, Chile. IEEE Computer Society, pp 2641–2649. Accessed from 7–13 Dec 2015
19. Krishna R, Zhu Y, Groth O, Johnson J, Hata K, Kravitz J, Chen S, Kalantidis Y, Li L-J, Shamma DA et al (2017) Visual genome: connecting language and vision using crowdsourced dense image nnotations. In: IJCV, pp 32–73
20. Fellbaum C (2010) Wordnet. In: Theory and applications of ontology: computer applications. Springer, pp 231–243
21. Wu Z, Palmer M (1994) Verb semantics and lexical selection. In: ACL, pp 133–138
22. Veličković P, Cucurull G, Casanova A, Liò P, Bengio Y (2018) Graph attention networks. In: ICLR, Adriana Romero
23. Huang B, Carley K (2019) Syntax-aware aspect level sentiment classification with graph attention networks. In: Proceedings of the 2019 conference on empirical methods in natural language processing and the 9th international joint conference on natural language processing (EMNLP-IJCNLP), Hong Kong, China, pp 5469–5477
24. Ma N, Mazumder S, Wang H, Liu B (2020) Entity-aware dependency-based deep graph attention network for comparative preference classification. In: Proceedings of the 58th annual meeting of the association for computational linguistics, pp 5782–5788
25. Guo Z, Zhang Y, Lu W (2019) Attention guided graph convolutional networks for relation extraction. In: Proceedings of the 57th annual meeting of the association for computational linguistics, Florence, Italy, pp 241–251
26. Chen D, Manning C (2014) A fast and accurate dependency parser using neural networks. In: Proceedings of the 2014 conference on empirical methods in natural language processing (EMNLP), Doha, Qatar, pp 740–750
27. Pennington J, Socher R, Manning CD (2014) Glove: global vectors for word representation. In: Empirical methods in natural language processing (EMNLP), pp 1532–1543
28. Devlin J, Chang M-W, Lee K, Toutanova K (2019) BERT: Pre-training of deep bidirectional transformers for language understanding. In: Proceedings of the 2019 conference of the North American chapter of the association for computational linguistics: human language technologies, Vol 1 (Long and Short Papers)
29. Hochreiter S, Schmidhuber J (1997) Long short-term memory. Neural Comput 9:1735–1780
30. Devlin J, Chang M-W, Lee K, Toutanova K (2019) BERT: pre-training of deep bidirectional transformers for language understanding. In: Proceedings of the 2019 conference of the North American chapter of the association for computational linguistics: human language technologies, Vol 1 (Long and Short Papers), Minneapolis, Minnesota, pp 4171–4186
31. Radford A, Wu J, Child R, Luan D, Amodei D, Sutskever I (2019) Language models are unsupervised multitask learners. OpenAI blog, p 9
32. Ruff L, Görnitz N, Deecke L, Siddiqui SA, Vandermeulen RA, Binder A, Müller E, Kloft M (2018) Deep one-class classification. In: Dy DG, Krause A (eds) Proceedings of the 35th international conference on machine learning, ICML 2018, Stockholmsmässan, Stockholm, Sweden,

Proceedings of machine learning research, vol 80. PMLR, pp 4390–4399. Accessed from 10–15 July 2018

33. Schlachter P, Liao Y, Yang B (2019) Deep one-class classification using intra-class splitting. In: 2019 IEEE data science workshop (DSW), pp 100–104

34. Perera P, Nallapati R, Xiang B (2019) OCGAN: one-class novelty detection using gans with constrained latent representations. In: IEEE conference on computer vision and pattern recognition, CVPR 2019, Long Beach, CA, USA. Computer Vision Foundation/IEEE, pp 2898–2906. Accessed from 16–20 June 2019

35. Kingma DP, Welling M (2014) Auto-encoding variational bayes. In: Bengio Y, LeCun Y (eds) 2nd international conference on learning representations, ICLR 2014, Banff, AB, Canada, Conference track proceedings. Accessed from 14–16 Apr 2014

36. Schölkopf B, Platt JC, Shawe-Taylor J, Smola A, Williamson RC (2001) Estimating the support of a high-dimensional distribution. In: Neural computation, pp 1443–1471

37. Liu FT, Ting KM, Zhou Z-H (2008) Isolation forest. In: Proceedings of the 2008 eighth IEEE international conference on data mining, pp 413–422

38. Hu W, Wang M, Qin Q, Ma J, Liu B (2020) HRN: a holistic approach to one class learning. In: Larochelle H, Ranzato M, Hadsell R, Balcan M-F, Lin H-T (eds) Advances in neural information processing systems 33: annual conference on neural information processing systems 2020, NeurIPS 2020 virtual. Accessed from 6–12 Dec 2020

39. Tack J, Mo S, Jeong J, Shin J (2020) Csi: Novelty detection via contrastive learning on distributionally shifted instances. Adv Neural Inf Process Syst 33:11839–11852

40. Conneau A, Kiela D, Schwenk H, Barrault L, Bordes A (2017) Supervised learning of universal sentence representations from natural language inference data. In: Proceedings of the 2017 conference on empirical methods in natural language processing Copenhagen, Denmark, pp 670–680

41. Bowman SR, Angeli G, Potts C, Manning CD (2015) A large annotated corpus for learning natural language inference. In: Proceedings of the 2015 conference on empirical methods in natural language processing, Lisbon, Portugal, pp 632–642

42. Chalapathy R, Chawla S (2019) Deep learning for anomaly detection: a survey. arXiv:1901.03407

43. Vrandečić D, Krötzsch M (2014) Wikidata: a free collaborative knowledgebase. Commun ACM 57(10):78–85

44. Mintz M, Bills S, Snow R, Jurafsky D (2009) Distant supervision for relation extraction without labeled data. In: Proceedings of the joint conference of the 47th annual meeting of the ACL and the 4th international joint conference on natural language processing of the AFNLP, pp 1003–1011

45. Wu S, He Y (2019) Enriching pre-trained language model with entity information for relation classification. In: Proceedings of the 28th ACM international conference on information and knowledge management, pp 2361–2364

46. Perera P, Oza P, Patel VM (2021) One-class classification: a survey. arXiv:2101.03064

47. Tax DMJ, Duin RPW (2004) Support vector data description. Mach Learn 45–66

Multi-agent Game Domain: Monopoly

7

T. Bonjour, M. Haliem, V. Aggarwal, M. Kejriwal and B. Bhargava

In the previous chapters we have looked at the visual domain and single-agent environments for action domains. In this chapter, we will apply the novelty framework to a multi-agent game environment. As an example of multi-agent game environment, we introduce a simulated version of the Monopoly board game. Monopoly is a multi-agent board game that involves four players taking turns by rolling a pair of unbiased dice and making decisions. The conventional Monopoly board consists of 40 square locations which include 22 real estate locations, 4 railroads, and 2 utility locations that players can buy, sell, or trade. Furthermore, there are squares that correspond to "Go," a jail location, card locations, and the free parking location. Figure 7.1 shows all assets, their corresponding purchase prices, and color.

T. Bonjour (✉) · M. Haliem · B. Bhargava
Department of Computer Science, Purdue University, 305 N. University Street,
West Lafayette, IN 47907-2107, USA
e-mail: tbonjour@purdue.edu

M. Haliem
e-mail: mwadea@purdue.edu

B. Bhargava
e-mail: bbshail@purdue.edu

V. Aggarwal
School of Industrial Engineering, Purdue University, 305 N. University Street,
West Lafayette, IN 47907-2107, USA
e-mail: vaneet@purdue.edu

M. Kejriwal
USC Information Sciences Institute, 4676 Admiralty Way, Marina Del Rey, CA 90292, USA
e-mail: kejriwal@isi.edu

© The Author(s), under exclusive license to Springer Nature Switzerland AG 2024   97
T. Boult and W. Scheirer (eds.), *A Unifying Framework for Formal Theories of Novelty*, Synthesis Lectures on Computer Vision,
https://doi.org/10.1007/978-3-031-33054-4_7

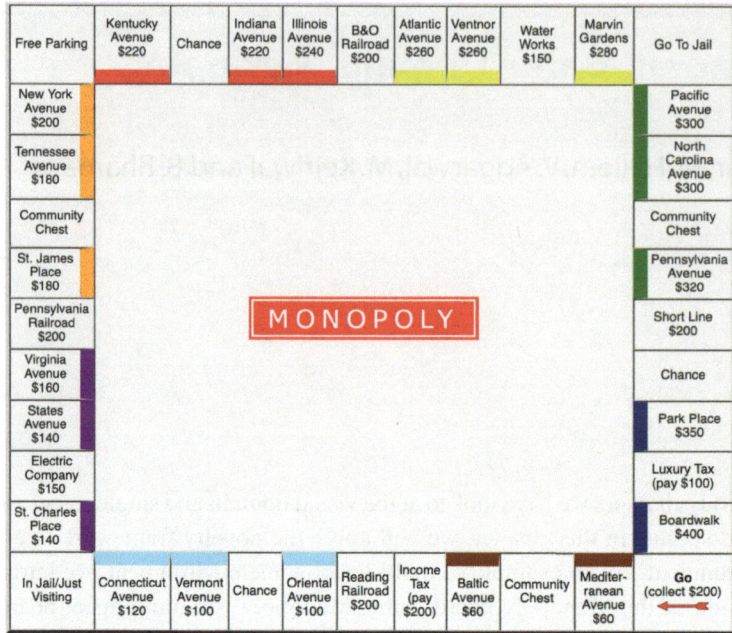

Fig. 7.1 Monopoly board [1]

7.1   Task Overview

We setup the Monopoly simulator to have one learning-based agent ($\alpha_{\mathcal{T}}$) and three fixed-policy agents. These constitute the four players in the game. The objective of the learning-based agent ($\alpha_{\mathcal{T}}$) is to learn winning strategies for Monopoly. Formally, the task $\mathcal{T}$ of the agent, $\alpha_{\mathcal{T}}$, is defined as: given the observation space $x_t \in \mathcal{O}$ at time t, select an action $a_t \in \mathcal{A}$ to maximize the overall reward resulting in a higher win rate.

7.2   Dissimilarity and Regret

Distance Metric: We can use distance metrics to define the dissimilarity measure in Monopoly. For instance, the observation space in Monopoly contains different location types that can be bought by a player: Real Estate, Railroads, and Utilities. Each location has multiple attributes: price and mortgage in the case of all three location types and price of building a house and rental rates in the case of a Real Estate location. A possible novelty could be a change in one or more such values in the Monopoly environment and a dissimilarity between two worlds can be calculated using a simple distance metric, e.g., Euclidean distance. Another example of novelty where a distance metric could be used as a dissimilarity measure would be a change in sequence of locations on a Monopoly board. An agent would possibly form an optimal policy based on the income that a particular location generates, so changing the prices or the relative location of a property will make the policy suboptimal in the novel world. Similar to the CartPole domain, we can measure the impact on performance of a novelty by training the agent in the pre-novelty environment, and then testing it in the post-novelty environment. However, unlike CartPole, which is a single agent environment, Monopoly is a multi-agent environment (with four players, in our case). When a novelty is introduced in a multi-agent domain, it would have an impact on all the agents, however, the extent of the impact may not be equal. In order to mitigate the effect of the novelty on the three fixed-policy agents, we would need to design agents that perform at the same (or similar) level in the post-novelty environments as the fixed agents perform in the pre-novelty environment.

Novelty Detection Using Replay Memory: We note that we can adapt the approach of change point detection on data consisting of experience tuples to detect the change in the environment dynamics. Novelty detection on experience tuples can, for instance, be used to detect the change in the set of rules that govern the environment dynamics, for instance: rolling a six with the die gives an additional turn rather than stopping there, or rolling a one moves three steps rather than one. These types of novelties will cause the transition probability to change, which will, in turn, disrupt the distribution across memory samples.

For this approach, the agent maintains a replay memory, which acts as an internal experience tuple that is associated with the agent. Through interaction with the World $\mathcal{W}$, the agent stores memory tuples m_t that consists of $[S_t, r_t, S_{t+1}]$, that is the current state, reward associated with action a_t, and next state. We note the next state is the result of mapping the current state and the agent's selected action at time t using the state transition function: $S_{t+1} = T(S_t, a_t)$. The samples from the replay memory can be analysed for context changes in batch mode or online mode. To compare two worlds ω and $\tilde{\omega}$, the agent analyzes its memory samples associated with ω versus those associated with $\tilde{\omega}$. To achieve this, we adopt the Online parametric Dirichlet Change-Point (ODCP) detection algorithm proposed in [2] to examine the data consisting of experience tuples. This algorithm transforms any discrete or continuous data into compositional data and utilizes Dirichlet parameter likelihood testing to detect change points. Although ODCP requires the multivariate data to be i.i.d. samples from a distribution, the justification in [3] explains the utilization of ODCP in the Markovian setting, where the data obtained does not consist of independent samples. The full algorithm for the Dirichlet Change-Point detection algorithm is shown in Algorithm 1. In our case, the input to the ODCP algorithm is the data consisting of memory tuples m_t that are stored by our agent. In this algorithm, the maximum likelihood estimation of Dirichlet distribution parameters is calculated for the cumulative data (stored through memory tuples) using Eq. 7.1 below,

$$\alpha_i^* = argmax_\alpha \log\Gamma\left(\sum_l \alpha_l\right) - \sum_l \log\Gamma(\alpha_l) + \sum_l ((\alpha_l - 1)(\log(\hat{x}_l))),$$

$$\text{where } \hat{x}_l = \frac{1}{T}\sum_i \log(x_{i_l}). \tag{7.1}$$

Then, the log likelihood given distribution Q_0 is calculated using Eq. 7.2 below,

$$LL(x_1 \ldots x_T, Q) = \sum_{i=1}^{T} \log(Q(x_i)), \ Q(x_i) = \frac{1}{B(\alpha)}\prod_{l=1}^{d} x_{i_l}^{\alpha_l - 1}, \text{ and } B(\alpha) = \frac{\prod_{l=1}^{d}\Gamma(\alpha_l)}{\Gamma(\sum_{l=1}^{d}\alpha_l)}$$

$$\text{where } d = |x_i| \text{ Dimensionality of } x_i, \ x_l \geq 0, \text{ and } \sum_{l=1}^{d} x_l = 1. \tag{7.2}$$

Then, at each time step t, that is seen as a potential change point, we split the data into two parts (prior and after this time step t), and we estimate the maximum likelihood for the cumulative un-split data (LL_0) as well as the sum of log likelihood for both partitions ($LL(t)$) using the equations above. Then, using the maximum log-likelihood LL^*, we obtain our dissimilarity measure of the two worlds $D_{o,T} = LL^* - LL_0$: that is the difference between the max log-likelihood and the log-likelihood of our un-split original data. Finally, if $D_{o,T} > \delta$, where δ is our dissimilarity threshold, then the data distribution between the two worlds is sufficiently different. In that case, the algorithm also returns the point in time T^* associated with the maximum log-likelihood LL^* to be the point in time when a novelty has occurred.

Agent Regret Function: Assume that we train an agent in the pre-novelty world w with three fixed-policy agents as opponents and obtain a near-optimal policy, π^*, for this environment. Given a state $s_t \in \mathcal{S}$, the learnt policy π^* guides the player to choose the best action $a_t \in \mathcal{A}$ at time step t. Since this is a multi-agent domain, we need to design the fixed policy opponents in the post-novelty world space, $\breve{w}$, such that they perform at par with the fixed policy agents in the pre-novelty world space, w. Then we define the regret as the difference in the win rates (Win_G) using the optimal policy π in the new world $\breve{w}$ and the win rates (Win_G) using the policy π^* in the new world $\breve{w}$ (where the win rate is defined as the percentage of game-wins in a tournament of n games). Thus, regret for the agent's policy (π^*) w.r.t. $\mathcal{T}$ is,

$$R_{o,T} = \max_\pi Win_G(\breve{w}, \pi) - Win_G(\breve{w}, \pi^*).$$

We note that the above definition of regret assumed a non-adaptive agent, that is learning policy π^* in the pre-novelty world, and is using the policy π^* in the post-novelty world. However, the agent can be adaptive, and keep learning or fine-tune the policies. In order to formulate the regret for the adaptive agent, let the world at time t (or tournament) be w_t, and π_t^* be the optimal policy for w_t (that optimizes the number of wins within the world w_t).

Algorithm 1 Dirichlet Change Point Detection Algorithm

1: **Input** Time Window $[1 \ldots T]$, Data $[x_1 \ldots x_T]$.
2: **Output** T^*: Change Point (if there is a change).
3: **procedure** DCP($[x_1 \ldots x_T]$)
4:     $Q_0 \leftarrow$ **Estimate** Drichlet Parameters for $[x_1 \ldots x_T]$ using Eq. 7.1
5:     $LL_0 \leftarrow$ **Estimate** Log-Likelihood for $[x_1 \ldots x_T]$ under Q_0 using Eq. 7.2
6:     $(T^*, LL^*) \leftarrow$ ESTIMATE_2WINDOW($[x_1 \ldots x_T]$)
7:     $D_{o,T} \leftarrow LL^* - LL_0$
8:     **if** $D_{o,T} > \delta$ **then**
9:         **Return** Change point at T^*.
10:    **else**
11:        No change, **Return**
12:    **end if**
13: **end procedure**
14: **procedure** ESTIMATE_2WINDOW($[x_1 \ldots x_T]$)
15:    **for** $t \in 1 \ldots T - 1$ **do**
16:        $Q_1 \leftarrow$ **Estimate** Drichlet Parameters for $[x_1 \ldots x_t]$ using Eq. 7.1
17:        $Q_2 \leftarrow$ **Estimate** Drichlet Parameters for $[x_{t+1} \ldots x_T]$ using Eq. 7.1
18:        $LL_t \leftarrow$ Log-Likelihood for $[x_1 \ldots x_t]$ under Q_1 + Log-Likelihood for $[x_{t+1} \ldots x_T]$ under Q_2 (Eq. 7.2)
19:    **end for**
20:    $LL^* \leftarrow max_{(t \in 1 \ldots T-1)} LL(t)$
21:    $T^* \leftarrow argmax_{(t \in 1 \ldots T-1)} LL(t)$
22:    **Return** (T^*, LL^*)
23: **end procedure**

As an example, assume w_t is a pre-novelty world (or tournament) and $\breve{w}$ is a post-novelty world. Let policy π_t be played by the agent in tournament t. Then, the regret for an adaptive policy over T tournaments can be defined as follows,

$$R^A_{o,T} = \sum_{t=1}^{T} \left(Win_G(w_t, \pi_t^*) - Win_G(w_t, \pi_t) \right).$$

7.3   Measurement and Observations

For this task we use the GNOME (Generating Novelty in Open-World Multi-agent Environments) Monopoly simulator [4] which is publicly available. In GNOME, the game board is initialized using a schema represented as a simple Javascript Object Notation (JSON) key-value data structure. Monopoly is a complex environment with multiple attributes, some of which are: current cash for each player, assets owned by each player, different location types, location sequence (relative positions of different locations on the board), income and luxury tax, and community and chance cards. To simplify the problem, we setup the agent task $\mathcal{T}$, as a Markov Decision Process (MDP) [5]. Since Monopoly is a fully observed environment, the world space $\mathcal{W}$ and observation space $\mathcal{O}$ are considered to be the same, and hence, there is no need for a perceptual operator for this task.

7.4   Novelty Types and Examples

Monopoly is one of the more sophisticated environments where novelties can take various forms. Since, in this case, the world space and observation space are considered to be the same, we can use observation or world novelty interchangeably. Below, we give some examples of observation novelty, agent novelty, and managed novelty.

As mentioned in the previous section, a change in the values of property prices is an example of observation novelty, where the dissimilarities are larger than zero, $\mathcal{D}_{o,\mathcal{T}}(\check{x}_t, x_t) > 0$.

An observation novelty occurs when the observation of the agent is sufficiently dissimilar from every past observation in the agent's stored replay memory. An example of such novelty could be: getting a six on the die gives the player an additional turn rather than stopping there, getting a one on the die moves three steps rather than one, or using biased, non-uniform die. These novelties will cause changes in the transition probability function, and thus can be detected with the Dirichlet distribution dissimilarity measure mentioned above, where $\mathcal{D}_{o,\mathcal{T}}(\check{x}_t, x_t) > \delta$.

An agent novelty is an observation-space state that the agent cannot map to any of its internal states. In Monopoly, inclusion of additional squares on the board or addition of new properties would be considered an agent novelty.

Taking regret into consideration, an example of a managed novelty that has a minimal impact on the agent performance could be: changing names of properties on the board, for instance, *St. Charles Place* is changed to *Charles Place*.

7.5   Experiments

In our setting, we consider a four player Monopoly game where we train a DQN agent against three other opponents (rule-based agents in our setting) with $\approx$ two-million games, i.e., episodes. The state space, action space, and reward for this agent are adopted from [6]. At each time step t, the DQN agent selects an action $a_t \in A(s_t)$ based on the current state of the environment $s_t \in S$, where S is the set of possible states and $A(s_t)$ is the finite set of possible actions in state s_t. We utilize the ϵ-greedy exploration policy [7] to select actions. After an action is executed, the agent receives a reward, $r_t \in R$, and state of the environment is updated to s_{t+1}. The transitions of the form (s_t, a_t, r_t, s_{t+1}) are stored in a cyclic buffer, known as the *replay buffer*. This buffer enables the agent to randomly sample from and train on prior observations.

Fig. 7.2 Performance of DQN
agent when varying sky-blue
mortgage values from 25 to
200 in Monopoly

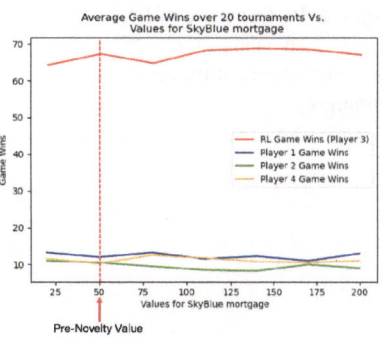

Fig. 7.3 Performance of DQN
agent when varying orange
prices per house values from 50
to 450 in Monopoly

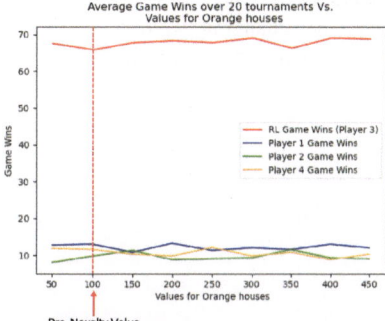

For evaluations, we test the performance of our pre-trained agent over 20 tournaments of
40 games each, i.e., total of 800 games, for two different types of novelties:

1. Change values for (mortgage or price per house) for a random color-group: in this experiment, we vary the values of either mortgage, prices, or price per house for some color
   group and investigate the performance, i.e., percentage of game wins, of our DQN agent
   pre- and post-novelty injection as compared to that of the three other opponent agents.
   We can observe, in Figs. 7.2 and 7.3, that the fluctuation in performance is very minimal,
   thus, this type of novelty does not affect the learning of the DQN agent and it is still able
   to perform well.
2. Change opponent player after a certain number of tournaments: in this experiment, we
   change one or more of the opponent players such that the strategy they follow is different
   from what our DQN agent had seen during training. We inject this novelty after a certain
   number of tournaments (here, after the beginning of the fifth tournament). Then we
   investigate the percentage of game wins of the DQN agent per tournament against that
   of the three other players. As shown in Figs. 7.4 and 7.5, this type of novelty greatly

Fig. 7.4 Performance of DQN
agent when changing one of
the three opponent players in
Monopoly

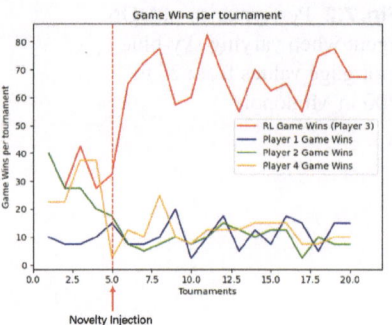

Fig. 7.5 Performance of DQN
agent when changing two of
the three opponent players in
Monopoly

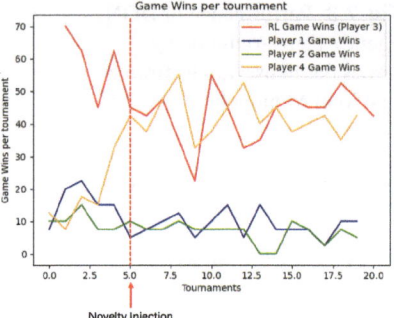

affects the performance of the DQN agent, which is evident by the large fluctuations in
the percentage of wins as the world dissimilarity increases. Thus, we can conclude that
that world-level dissimilarity is correlated with world-level performance.

7.6   Conclusions

Monopoly is a complex multi-agent environment, and the agent's decision-making is
designed to mimic many elements of the real world like skill, luck, and modeling of opposing players' strategies. It provided a very good testing bed for the novelty framework. As
discussed earlier, the dissimilarity metrics in Monopoly depend on the type of novelty being
introduced. For environment novelties, where we changed an attribute of the environment,
for instance, the rent, a distance metric may suffice, but for more complex novelties, changing of opponent's strategy, we needed a more complex metric like the difference between
the distribution of transitions. Not all novelties affected the agent similarly as shown in our
experiments. In this work, we implemented the three opposing players as rule-based. In
the future, we plan to analyze the behavior and apply the novelty framework to situations
where more than one player is learning-based. In our experiments, we showed the effect of

the introduction of a single novelty on the performance of the agent. Another area worth exploring in the future is to analyze the effect of composite novelties where we introduce multiple novelties at the same time.

References

1. Wikipedia. Template: monopoly board layout. https://en.wikipedia.org/wiki/Template: Monopoly_board_layout
2. Singh N, Dayama P, Pandit V et al (2019) Change point detection for compositional multivariate data. arXiv:1901.04935
3. Padakandla S, Bhatnagar S et al (2019) Reinforcement learning in non-stationary environments. arXiv:1905.03970
4. Kejriwal Mayank, Thomas Shilpa (2021) A multi-agent simulator for generating novelty in monopoly. Simul Modell Pract Theory 112:102364
5. Ash RB, Bishop RL (1972) Monopoly as a markov process. Math Mag 45(1):26–29
6. Bonjour T, Haliem M, Alsalem A, Thomas S, Li H, Aggarwal V, Kejriwal M, Bhargava B (2022) Decision making in monopoly using a hybrid deep reinforcement learning approach. IEEE Trans Emerg Top Comput Intell
7. Mnih V, Kavukcuoglu K, Silver D, Rusu AA, Veness J, Bellemare MG, Graves A, Riedmiller M, Fidjeland AK, Ostrovski G et al (2015) Human-level control through deep reinforcement learning. Nature 518(7540):529–533

Concluding Thoughts

8

T. Boult and W. Scheirer

We see three primary contributions of this formalization of novelty that will spur further research. First, formalization forces one to specify (or intentionally disregard) the required items in the theory. This can lead to insights about the problem and fill in knowledge gaps. For example, when applying the theory to the CartPole domain, numerous unanticipated issues were highlighted, new predictions made, and new experiments validated the new insights.

Second, formalization provides a common language to define and compare models of novelty across problems. The precision of terms reduces confusion, while the flexibility allows it to be applied to a wide range of problems.

Third, the formalization allows one to make predictions about where or why experiments incorporating some form of novelty might run into difficulties. For example, when the world-level and perceptual-level dissimilarity assessments disagree, we predict novelty problems will be more difficult. One example of difficulty is world-disparity using variables not represented in perceptual space. Another is when there are many possible world labels, but the input is only assigned one label that is used for assessing world-level dissimilarity. In this case, the theory predicts a greater difficulty with such novelty, especially if the assigned label is associated with a physically smaller aspect of the observation.

Biological intelligence has a remarkable capacity to generalize novel inputs with ease, yet artificial agents continue to struggle with this behavior. It is our hope that the adoption and use of the framework proposed here leads to the development of more effective solutions for novelty management, and to make agents more robust to novel changes in their world.

T. Boult (✉)
University of Colorado Colorado Springs, Colorado Springs, CO, USA
e-mail: tboult@vast.uccs.edu

W. Scheirer
University of Notre Dame, Notre Dame, IN, USA

© The Author(s), under exclusive license to Springer Nature Switzerland AG 2024   107
T. Boult and W. Scheirer (eds.), *A Unifying Framework for Formal Theories of Novelty*, Synthesis Lectures on Computer Vision,
https://doi.org/10.1007/978-3-031-33054-4_8

By formalizing seven different activity and perceptual domains using our novelty frame-work, we gained insights into what are meaningful "novelty" problems for the various tasks associated with the domains. We showed how to develop better measures to predict when novelty would be easy or hard to manage or detect. In line with this, our team of researchers continue to refine this theory and apply it to other problem domains.

An important future direction is further theoretical development that is informed by disciplines outside of computer science. Three things come together as an agent explores the world: the environment, sensory perception, and activity. An agent senses the environment and takes action based on the information that is sensed. This means novelty can be relative to the environment, the perception system, or the agent. Novelty can also cut across these three categories. Crucially, a comprehensive theory of novelty can benefit from prior perspectives on these areas. Let us briefly review the implications that this work has on any theory of novelty for AI.

First, what does it take for something in the environment to be different? This is a theoretical question that has been raised in the study of the philosophy of mind. David Chalmers [1] has argued that novelty always takes one of two forms: *strong emergence* or *weak emergence*. In the case of strong emergence, novelty is irreducible and leads to downward causation. Thus, it is novel to the whole of something instead of to the parts. In the case of weak emergence, properties emerge from a system that are unexpected given the rules that govern its lower levels. Thus, this view is reductive. Both of these concepts rely on an understanding of the levels of abstraction, with more empirical support existing for the latter but the possibility exists for both. Thus, a comprehensive theory of novelty must account for all types of novelty emanating from the environment. More important from an AI perspective, however, is the agent's reaction when encountering it in any form.

While novelty can exist external to an agent, it is experienced through sensory perception. The same is true of biological organisms in a real environment. In human psychology, the visual system is the best studied sensory perception system, and novel stimuli are routinely used to probe it. For instance, various experiments have been formulated to test the human mechanisms of visual attention, largely through tasks involving the ability of a subject to discriminate oddball objects [2–4]. Psychophysical measurements, such as measured accuracy and reaction time, reflect the human ability to adapt to changes to the environment, which has obvious evolutionary advantages. This is captured in theoretical work on how sensory perception works in the brain [5, 6].

From a theoretical standpoint, novelty assignment in the perceptual regime is the output of the decision making process of the perceptual system—not the environment. Thus, a decision that a stimulus is novel can be matched to the ground-truth of the environment but that is not always the case. Humans experience new things all of the time. Thus, a decision that a stimulus is novel might mean that it is new to the sensory system experiencing it (especially true in development), but perhaps not new to the environment in which it is found. In other cases, the decision might simply be a mistake. Sensory perception in nature is a noisy process, especially in the presence of other environmental factors.

In the most consequential link of the process, the decision of whether or not something is novel from the perceptual system leads to specific agent behaviors. As in the previous two aspects, this process has a direct analog in biology. Activity driven by novelty in biological organisms can be divided into neophobia [7] and neophilia [8]. In behaviors associated with neophobia, there is fear or avoidance of novelty. In neophilia, there is attraction or preference for novelty. A bridge from these concepts to AI is the reinforcement learning paradigm. Gershman and Niv explored the problem of assessing the value of an activity that has not been tried before [9]. Theoretically, this is cast as the generalization of previous experience with one set of options to a novel activity in a Bayesian reinforcement learning context. However, it is also constrained, in that out-of-the-box, reinforcement learning lacks effective novelty detection and characterization abilities.

In this book, we presented a framework for generating theories of novelty, as they would apply to learning-based agents operating in an open-world setting where incremental learning is possible. We presented a formulation in which one can define a wide range of novelty problems. We contend any constructive or generative theory of novelty [10] must be incomplete because the construction or generation of defined worlds, states, and any enumerable set of transformations between them form, by definition, a closed-world. We note, however, that one can embed a constructive model within our definition. For example, SARS-Cov-2 is similar to SARS-Cov-1 but is a rearrangement of existing virus components from a closed-world. However, it is called a novel coronavirus. The key to extending the proposed framework is to gather insights from other fields and see how they fit. We invite researchers interested in the novelty problem to follow this path.

References

1. Chalmers DJ (2006) Strong and weak emergence. The re-emergence of emergence. Oxford University Press Oxford, pp 244–256
2. Nakayama Ken, Mackeben Manfred (1989) Sustained and transient components of focal visual attention. Vis Res 29(11):1631–1647
3. Bravo MJ, Nakayama K (1992) The role of attention in different visual-search tasks. Percept Psychophys 51(5):465–472
4. Joseph JS, Chun MM, Nakayama K (1997) Attentional requirements in a 'preattentive' feature search task. Nature 387(6635):805–807
5. Marr D (1970) A theory for cerebral neocortex. Proc Roy Soc Lond Ser B. Biol Sci 176(1043):161–234
6. Fahle M, Poggio T, Poggio TA et al (2002) Perceptual learning. MIT Press
7. Corey DT (1978) The determinants of exploration and neophobia. Neurosci Biobehav Rev 2(4):235–253

8. Hughes RN (2007) Neotic preferences in laboratory rodents: issues, assessment and substrates. Neurosci Biobehav Rev 31(3):441–464

9. Gershman SJ, Niv Y (2015) Novelty and inductive generalization in human reinforcement learning. Top Cognit Sci 7(3):391–415

10. Langley P (2020) Open-world learning for radically autonomous agents. In: AAAI. JSTOR, pp 13539–13543